AN ATLAS OF IMMUNOFLUORESCENCE IN CULTURED CELLS

AN ATLAS OF IMMUNOFLUORESCENCE IN CULTURED CELLS

Mark C. Willingham

Ira Pastan

Laboratory of Molecular Biology
National Cancer Institute
National Institutes of Health
Bethesda, Maryland

1985

ACADEMIC PRESS, INC.
(Harcourt Brace Jovanovich, Publishers)

Orlando   San Diego   New York   London
Toronto   Montreal   Sydney   Tokyo

ACADEMIC PRESS, INC.
Orlando, Florida 32887

United Kingdom Edition published by
ACADEMIC PRESS INC. (LONDON) LTD.
24–28 Oval Road, London NW1 7DX

Library of Congress Cataloging in Publication Data

Willingham, Mark C.
    An atlas of immunofluorescence in cultured cells.

    Bibliography: p.
    Includes index.
    1. Immunofluorescence–Atlases.   2. Fluorescent
antibody technique–Atlases.   3. Cell culture–Atlases.
4. Cells–Atlases.   I. Pastan, Ira H.   II. Title.
[DNLM: 1. Fluorescent Antibody Technic–atlases.
QW 517 W733a]
QR187.148W56   1985   574.87'028   85-6047
ISBN 0–12–757030–6 (alk. paper)

PRINTED IN THE UNITED STATES OF AMERICA

85 86 87 88   9 8 7 6 5 4 3 2 1

To

Thomas A. Willingham, Jr., and Eleanor C. Willingham

and to

Miriam Pastan and Harvey Pastan

CONTENTS

PREFACE

Immunofluorescence has become a standard method for the study of cellular organization. Thus, fluorescent images appear in almost every issue of journals devoted to cell biology. However, to our knowledge no collection of representative images produced by antibodies to different cell proteins in cultured cells has been published. The purpose of this atlas is to illustrate the location of various cellular proteins using fluorescently labeled antibodies. The examples shown are from our own studies. In many cases, the proteins are found associated with specific organelles, and these patterns can be used to identify the organelles. Thus, the images included in this atlas provide a general outline of the types of patterns that exist for many cellular organelles. We have not attempted to include every possible pattern that might be seen. Such an endeavor would require an extensive volume. Also, we have not included many different antigens that are found in the same sites and, therefore, give the same pattern. Rather, we have selected representative examples. Also, patterns observed in unusual cell types, especially with cytoskeletal antigens, are not included. We have included only those more common patterns that cytoskeletal elements assume in frequently used fibroblastic and epithelioid cultured cell lines.

It is a general observation that cell types that are highly adhesive to their substratum spread out and produce much clearer images of intracellular organelles and structures than rounded thick cells. Thus, many of the images included in this atlas employ flattened cell types. Cells that are more rounded make clear delineation of intracellular structures more difficult. If plasma membrane-related antigens are of interest, then more rounded cells are useful because of the ease of seeing plasma membrane labeling at the vertical edge of a rounded cell.

Immunofluorescence has become a useful tool in the study or selection of monoclonal antibodies. Screening can be accomplished using cultured cell lines, and the patterns seen by immunofluorescence can provide important clues as to the usefulness of a particular clone. For example, an investigator may wish to save only those clones that are directed at cell surface components, cytoskeletal elements, or other cellular organelles. Immunofluorescence provides a quick and easy technique to classify these antibodies on the basis of morphology.

Immunofluorescence also has other very important advantages. Antigens that are present in only a small percentage of cells in the population may be readily identified. In comparison with other light microscopic immunocytochemical techniques, such as peroxidase methods, immunofluorescence can detect much smaller sized sites of antigen location. Objects too small to be seen by techniques that depend on light refraction for detection can be detected as a point source of light by their fluorescence emission. Thus, fluorescence can be used to clearly resolve objects as small as individual microtubules or coated pits. For antigens present in most cells, but actually present only as a single structure in each cell, such as a nematin fiber or a primary cilium (see text), immunofluorescence offers the only simple technique for their detection.

Chapter 1 on methodology describes some of our experiences with different techniques and emphasizes the simplest approach for general screening procedures. The choice of a technique is often a matter of personal preference and the need for special applications. We have tried to provide sufficient information to help investigators unfamiliar with fluorescence make such a choice.

Mark C. Willingham
Ira Pastan

ACKNOWLEDGMENTS

The authors wish to thank Drs. Klaus Hedman and Karen Goldenthal for reviewing the manuscript, and Angelina Rutherford, Susan Yamada, and Maria Gallo for expert technical assistance. We also thank the following investigators for their generous gifts of antibodies used in the preparation of this atlas: Drs. M. E. Furth, K. P. Hedman, G. W. H. Jay, C. B. Klee, E. Racker, N. D. Richert, G. G. Sahagian, E. M. Scolnick, and, especially, Dr. J. T. August.

AN ATLAS OF IMMUNOFLUORESCENCE IN CULTURED CELLS

1

IMMUNOFLUORESCENCE TECHNIQUES

1

A.   INTRODUCTION

This chapter outlines the methods and procedures that we have found practical for immunofluorescence using cultured cells. These techniques are designed to allow large numbers of
samples to be handled quickly and reproducibly. In addition, the sensitivity and resolution of
these techniques is high enough to allow evaluation of images when only low concentrations of
antibodies are available. For convenience we routinely use plastic dishes. However, the clarity
of some very bright images is not necessarily optimal with these dishes, because plastic dishes
may introduce a slight background due to the autofluorescence of plastic that is absent with
samples mounted on glass coverslips. A very bright image of microtubules would be somewhat
clearer on a glass coverslip; on plastic the autofluorescence of the surrounding substrate keeps
the background from being completely black. Nevertheless, plastic dishes are very convenient
and yield readily interpretable and clear images.

The structures that are visualized with images of monolayers of cultured cells are characteristically displayed in a plane parallel to the substratum (Fig. 1). It is important to remember
when making structural interpretations that these images are representations of the three-
dimensional shape of these cells. A drawing is shown in Figure 2 that schematically represents
the relationships of intracellular structures in a plane perpendicular to the substratum. Some of
the elements of this drawing emphasize important points to keep in mind when interpreting
cultured cell images, particularly of very flattened cells. A flat fibroblastic cell can occupy an
area on the substratum of 3000–15,000 μm^2 and be only 0.2–5.0 μm thick. Thus, such a cell is
very asymmetric. As one observes structures such as microtubules coursing through the
cytoplasm from the cell center to the cell margin, one must constantly keep in mind this third
dimension. For example, the microtubule is usually within a few microns of the plasma
membrane, and at the thinner edges at the cell margin it is less than 1.0 μm from the surface.
Yet, the microtubule may be 50–100 μm from a lateral cell margin. While there is plasma
membrane at the cell margin, the majority of the plasma membrane lies over and under the
image of the rest of the cell. This confusion has led some to refer to the path of microtubules
from the cytocenter to the cell margin as reflecting a general path from cell center to cell

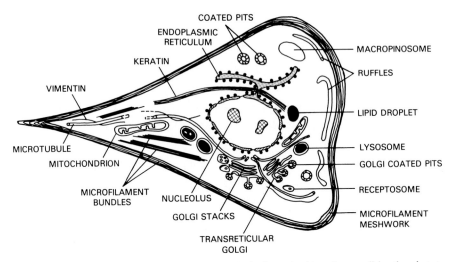

Figure 1.   A diagram of organelles and structures in a cultured cell visualized in a plane parallel to the substratum.

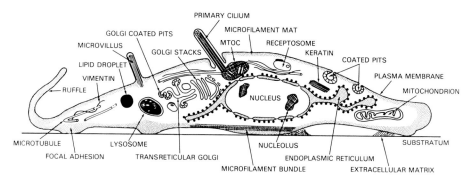

Figure 2. A diagram of organelles and structures in a cultured cell visualized in a plane perpendicular to the substratum.

surface; it should be clear that this path has nothing to do with the relationship to the cell surface, but only to the cell margin. Similarly, a point over the central perinuclear "Golgi" region allows any structure within the cell in this area to be no more than 2.5 μm from the cell surface (above or below the cell). It should be clear that some cultured cells are very asymmetric and the lateral cell margin and the cell surface have no relationship to each other.

The membranous structures illustrated in Figures 1 and 2 include many intracellular elements that interact with each other both physically and functionally. Thus, elements of the plasma membrane, endocytic vesicles, lysosomes, Golgi, and endoplasmic reticulum–nuclear envelope are all functionally interconnected in some very specific ways. However, the one membranous organelle that does not regularly interact with, and stays physically disconnected from, other intracellular membranes is the mitochondrion. This is in keeping with its probable origin during evolution as a symbiotic parasite.

An important point in all of immunocytochemistry is the significance of negative results. Sequestration of antigens can occur in very dense structures in cells, either naturally because of their structure or artificially induced by fixation. This problem is more serious in preservation for electron microscopic immunocytochemistry, where fixatives that produce better preservation also tend to make some cellular sites more compact and less permeable to antibodies. However, the problem also exists for immunofluorescence; one example is the almost complete impermeability of the midbody in the intercellular bridge that forms after mitosis. With the use of monoclonal antibodies, accessibility problems have become more common. Monoclonals recognize single determinants, and many of these may be in proteins whose structures are greatly affected by chemical fixatives. Thus, blind screening for an important antigen should include primary fixatives that are likely to leave such antigenic sites accessible. Such choices are usually only educated guesses based on the suspected structure of the antigen. A similar problem related to single antigenic determinants may also be encountered with the use of synthetic peptides as antigens. If an antigen on the cell surface is being studied, no fixation may be required.

Even the permeabilization agent may affect these results. Some proteins in membranes may not have accessible cross-linking sites such as amino groups in their structure. Thus, when a permeabilization reagent such as a detergent is used, the protein will be solubilized and removed from the fixed cell because it is not cross-linked in place. We have encountered three such membrane proteins. The solution to this type of problem was to use saponin as a permeabilization reagent. Saponin does not remove all phospholipids; it mainly extracts cholesterol, so that even if the antigen is not fixed in place by chemical cross-links, the membrane proteins will not be completely removed from the phospholipid of the membrane. In other cases, detergents may interact with proteins to change their structure so that antigenic sites are rendered inaccessible. Such conformational changes may become more of a problem because of the high selectivity of monoclonal antibodies.

It is very important to use the proper optics for immunofluorescence. Fluorescence techniques can visualize very small numbers of molecules inside cells, but only if the optics are of sufficient resolving power and sensitivity. The major element in such systems is the objective lens. For immunofluorescence of single cultured cells, an objective lens of high numerical aperture is essential. We commonly use planapochromat objectives with numerical apertures greater than 1.0. This includes 40× oil (N.A. 1.0) and 63× oil (N.A. 1.4) objectives.

Epifluorescence is far superior to substage dark-field fluorescence for excitation. This is particularly useful when using thick supporting surfaces such as plastic dishes. In addition, epi-illumination allows the simultaneous use of phase contrast, Nomarski, or Hoffman modulation optics so that a bright-field image can be recorded from the same field as the fluorescence image. While one can use substage illumination dark field with fluorescence filters and dark field using visible light for a comparative image, the thinness of the support material that is needed for dark field makes this difficult, unless one uses only standard microscope slides as a support.

Coverslip cultures have been commonly used for immunofluorescence with excellent technical results, but handling coverslips routinely is difficult and the mixing of small amounts of solution over the coverslip surface is not always dependable. Further, when coverslips are inverted onto a mounting medium on a slide, the coverslip must be immobilized to prevent motion during photography. Mounting media that solidify are available for this purpose, but the overall amount of work involved is much greater than the simplified plastic dish technique we use for routine screening. Particularly for beginners, coverslip techniques are difficult to master quickly.

In the last decade, cell culture experiments have led more and more investigators to use immunofluorescence. The previous decade had seen a decline in the use of fluorescence techniques in favor of peroxidase methods. The reason for this was that most of these earlier studies utilized tissue sections rather than cultured cells. For tissue sections, particularly those processed using paraffin embedding, the autofluorescence inherent in the sections makes fluorescence a less useful technique. For immunofluorescence with tissue, one must employ unfixed or lightly fixed frozen sections with a resultant dramatic loss in preservation of structure. For the most part, tissue section immunocytochemistry deals with the detection of antigens in cell types, rather than in individual organelles in single cells, and high levels of resolution are not required. Because peroxidase techniques generate their images using refraction, very small objects below the limit of refractile resolution will not be resolved. Fluorescence, on the other hand, generates images by light emission and can detect objects much smaller than the refractile limits of resolution. Thus, when investigators began to study individual single cells in culture and individual organelles in those cells, fluorescence methods had clear advantages over peroxidase or other catalytic labels. As a result, the method of choice for tissue sections is peroxidase, whereas the method of choice for single cultured cells is fluorescence.

More recently, interest has been generated for methods of coupling immunoglobulin reagents to markers using techniques other than the usual antiglobulin second-step conjugates of indirect immunofluorescence. One such method is the use of the avidin–biotin system. For some applications this approach has unique advantages. In general, however, the use of these reagents was inspired by the lack of availability of affinity-purified second-step conjugates of antiglobulins. This problem has generally been resolved with the commercial availability of high-quality affinity-purified fluorescence conjugates of antiglobulins. Some special applications may benefit from the avidin–biotin system. We routinely perform indirect immunofluorescence using affinity-purified antiglobulin conjugates.

Immunofluorescence is not intended generally as a quantitative technique. With special equipment it is possible to measure fluorescence intensities quantitatively, but the interpretation of this information has to be based on careful assessment of the morphology of the structures involved and the accessibility of the antigenic sites of interest. While one can gain a "feeling" about the quantitative aspects of a fluorescence image, it is not our recommendation to place a great deal of faith in such impressions. For freely accessible sites, such as on cell surfaces, careful comparisons of fluorescence images, particularly using equal time photographic exposures, can discern differences of a factor of 2 or so; however, this should not be depended on as an accurate indication of the number of antigenic sites.

For cell surface antigens, measurements using fluorescence-activated cell sorters can yield high sensitivity and quantitative results. These methods usually utilize living cells in suspension. For antigens that are in intracellular locations, however, these techniques are not readily applicable. Immunofluorescence on cell surfaces can employ either immobilized attached cells or living cells in suspension, processed using rapid centrifugations. These experiments are relatively simple in their interpretation and are not the subject of the images presented in this atlas.

Immunofluorescence is also a very useful technique as a primary screening method for the fixation conditions or other control experiments in preparation for pre-embedding electron microscopic immunocytochemistry. It is usually a safe assumption that antigens that are not detectable by immunofluorescence will be uninterpretable by electron microscopy. With some techniques, the same cell can be viewed by immunofluorescence and later by electron microscopy, a significant advantage for experiments using methods such as single-cell microinjection.

B. FIXATION AND PERMEABILIZATION

In general, immunofluorescence is usually performed after a primary fixation step. An exception would be the labeling of living cells, either at 4°C to prevent endocytosis of surface label or at room temperature for detection of surface antigens. These cells are usually fixed after the antibody incubations to preserve them for mounting under a coverslip. For intracellular antigens, cells require primary fixation to immobilize them on the substratum and to prepare them for treatment with a permeabilizing reagent for intracellular localization. The routine primary fixative we use is formaldehyde, usually 3.7% (a 1 : 10 dilution of stock 37% formalin). In some instances freshly made paraformaldehyde has been used, but it is not routinely necessary to prepare formaldehyde in this manner. The primary formaldehyde is made in an isotonic saline solution, usually phosphate-buffered saline (PBS). If calcium and magnesium are employed in the buffer, the cells may remain attached better during fixation. On the other hand, microtubules do not preserve as well because of their calcium sensitivity. Cultured cells are immediately accessible to the fixative solution, so that an extended fixation time is not necessary to allow permeation. For convenience, fixation is performed at room temperature; usually 10–15 minutes is sufficient for preservation to be accomplished. The fixed cells are immobilized on the substratum by this fixation. The cells should *never* be allowed to dry, since this severely affects morphologic preservation! Other primary fixatives include denaturing organic solvents, such as acetone or methanol. The morphologic preservation achieved with these primary fixatives is considerably worse than with formaldehyde.

After the primary fixation step, the cell membranes are still present and restrict the entry of large proteins such as antibodies. Organic solvents can permeabilize cells at this step, and cold methanol or a solution of acetone in H_2O is suitable for this purpose. For plastic dishes, a mixture of 80% acetone/20% H_2O permeabilizes the membranes but does not dissolve the plastic dish. The acetone should not be prepared in PBS since it produces a precipitate. Again, the cells should never be allowed to dry. In addition to their permeabilization effects, organic solvents denature and precipitate proteins in the cell. Even though these organic solvents work to permeabilize cells, a better alternative is to use detergents, which solubilize the phospholipid membranes with less apparent denaturation of the fixed proteins. Of the detergents in common use, 0.1% Triton X-100 is a good choice. This permeabilization step need only last 2–5 minutes at 23°C.

When cells fixed and permeabilized using formaldehyde and Triton X-100 are examined using electron microscopy, it is apparent that much of the membrane structure is completely gone and the cytoplasmic morphology is poor. An alternative to completely solubilizing the membranes with Triton X-100 is to use a more selective detergentlike molecule, saponin. This substance solubilizes mainly cholesterol, leaving much of the membrane structure intact. The unusual thing about the use of saponin is that a morphologically discernible membrane remains. If one incubates a saponin-treated cell with proteins in solution in the absence of saponin, the membrane may appear completely impermeable. The reason may be that saponin has intercalated in the membrane structure in place of cholesterol and the membrane is still a continuous structure. To use saponin as a permeabilization reagent, it must be kept constantly present in the antibody incubations and washes. Since this has no effect on the antigen–antibody interaction or on the use of fluorochromes, this inclusion of saponin has no disadvantages. Note, however, that saponin may not permeabilize the nuclear envelope or mitochondrial membranes adequately for the detection of antigens in these organelles.

On occasion, the preservation using formaldehyde is not adequate for the structures of interest. One such case is the preservation of microtubules. When these structures are visualized using anti-tubulin antibody in formaldehye or methanol-fixed cells, they show irregularities in their outlines. A more adequate fixative is glutaraldehyde. However, glutaraldehyde induces some autofluorescence, increases background binding of antibodies because of residual aldehyde groups, and renders much of the cytoplasm too dense to allow antibodies to penetrate the fixed cell matrix. A solution to this problem is the use of a sodium borohydride treatment of the cell after the glutaraldehyde fixation step. This is accomplished by making a fresh solution of sodium borohydride in PBS (0.5 mg/ml) and quickly washing the fixed cells for a minute or two with this solution. The residual borohydride is then removed by washing in PBS. This treatment, introduced by Weber, Rathke, and Osborn in 1978 (4), provides exceptional preservation of microtubules. Unfortunately, this method is not universally applicable to all other sites in the cell because of the increase in matrix density induced by the fixative (5).

The method of fixation we employ for routine screening of antibodies for general use is as follows: 3.7% formaldehyde for 10 minutes at room temperature in PBS, followed by washes in PBS and incubations in antibodies in the continuous presence of 0.1% saponin in all subsequent incubations and washes (see Appendix). An alternative to the use of saponin is a single treatment with 0.1% Triton X-100 in PBS for 5 minutes at room temperature.

If only cell surface antigens are to be examined, then primary fixation in formaldehyde alone is adequate without permeabilization. One should be aware of a common artifact that occurs in this circumstance: small blisters can form on the cell surface which occasionally are disrupted, yielding a small hole in the surface. This leads to a variable pattern of discrete spots of nonspecific intracellular label.

C.   MECHANICS OF PROCESSING

For immunofluorescence detection of cell surface antigens, one can simply suspend cells in antibody solutions, separate and wash these cells using gentle centrifugation (or 15,000 g for 30 seconds), and place a drop of these suspended cells on a glass slide under a coverslip for observation. Such centrifugation steps do not work well after fixation. For intracellular antigens the problems are quite different.

Cultured cells have commonly been processed for immunofluorescence using coverslip cultures. Small sterile coverslips are placed in a culture vessel and cells are allowed to settle and attach. At some later time, the coverslips are removed and dipped into fixative solutions, often an organic solvent such as methanol or 100% acetone, and often followed by air drying. Then, small volumes ($<$10 μl) of antibody solutions are added to a limited area of each coverslip and placed in a moist chamber at 37°C, followed by washing each coverslip by dipping in wash solutions. The coverslip is held for this wash step using fine tweezers or special coverslip carriers. Alternatively, coverslips can be incubated with larger volumes of solutions by placing them in the bottom of a tissue culture dish. Finally, each coverslip is placed with the cells facing down on a glass slide in a solution that contains a mounting medium that will later solidify (e.g., Gelvatol), or the edges of the coverslip must be glued in place using nail polish. The glass slide is then viewed using a standard slide carrier stage on a fluorescence microscope. The advantages of this technique are mainly that the background fluorescence of the glass is very low and small volumes of antibody solutions (usually discarded) are used. But there are a number of disadvantages to the use of coverslips. For the highest numerical aperture objectives one must use #1 coverslips, which are very fragile. Handling these small, thin pieces of glass is very difficult for the beginner to accomplish without breaking the coverslip. Some cultured cells adhere relatively poorly to glass; they usually adhere much better to tissue culture grade plastic dishes. As a result, many cells may be lost during washes using coverslips. The coverslips must be sterilized and the cells must be planted ahead of time. When small volumes of antibody solutions are used, they are subject to drying, so that humidified chambers are necessary. The small volume of antibody usually cannot be recovered after use. The drop of antibody solution on the coverslip often has poor mixing qualities, such that different areas of the coverslip are exposed to different concentrations of antibody in an uncontrolled fashion. The mounting medium must be a special water-miscible, solidifiable material which must be allowed to dry before viewing the specimen (to prevent drift during prolonged photographic exposures). Alternatively, the edges of the coverslip must be immobilized with nail polish and allowed to dry, a rather tedious process. This immobilization is important if any photography is contemplated, since the cells are attached to the coverslip and not to the glass slide. Therefore, the entire process takes a long time and only very adherent cells can easily be viewed in this way.

We prefer an alternative technique that is simpler and much more reproducible for large numbers of samples. We routinely prepare cultures in 35-mm tissue culture dishes. These may be used for propagation, for immunofluorescence, or for other purposes. The cells adhere well to these dishes. The volume of an incubation solution in such a dish necessary to guarantee that no drying occurs is between 0.5 and 1.0 ml. The incubations are carried out at room temperature on the bench top without the need for a humidified chamber. We usually use a slowly moving, small, tilting platform to gently mix the solutions in the dishes. This platform is a modified tilting device, made for mixing blood in tubes, in which the tilt angle has been reduced internally by moving the drive pin to a narrower radius. The final tilt angle is $\pm 5°$ and the period of the tilting motion is 3.5 seconds.

Cells are fixed and permeabilized directly in the dishes, and washing is performed simply by pouring the solution in the dish out and pouring wash solution in from a large bottle. The antibody incubation solutions have a volume of 1 ml but are recovered from the dish using a Pasteur pipette and are saved and frozen for reuse (see Section D). At the end of the incubations, the cells are mounted under a #1, circular 25-mm-diameter coverslip by adding a drop of buffered glycerol solution to the center of the dish and dropping the cleaned coverslip on top of the glycerol. The residual buffer that is displaced is absorbed from the edge of the coverslip with a tissue, and the dishes are then placed on the stage of an upright microscope. The stage used can be specially designed to hold a dish, or a hole can be made in a plate of any size (if phase-contrast substage illumination is also used), or a flat opaque plate can be used for epifluorescence alone. A drop of immersion oil is placed on top of the coverslip in the dish, and an objective is then rotated into position in the middle of the dish (see Fig. 3). These preparations can be viewed and then stored by replacing the cover of the dish for later viewing. There is no need to have an immobilizing medium under the coverslip, since the cells are

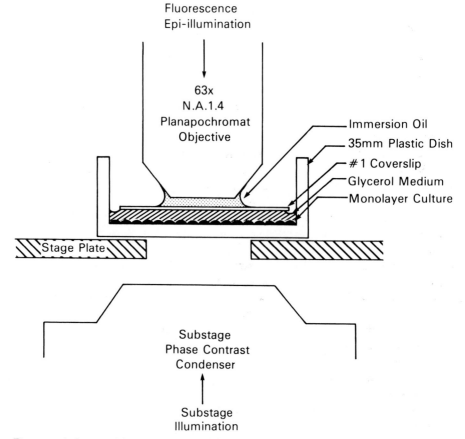

Figure 3. A diagram of the mounting method for immunofluorescence with cells grown directly on a tissue culture dish.

attached to the dish, not the coverslip. At a later time, the coverslip can be removed and the cells rehydrated with PBS and processed with another antibody for another marker, since they have never been in a nonaqueous environment. It is an easy task to process 20 or 30 samples at a time with this method, and no special skills are required to handle the dishes. Similar preparations can also be used for electron microscopy using *in situ* embedding methods.

This method has only slight disadvantages. Some dishes have inherent autofluorescence, especially in the fluorescein spectrum. This is one reason we usually use rhodamine as a label. However, fluorescein can still be used with this method if the illuminated field is reduced to a smaller portion of the image; this is usually accomplished by closing down the excitation aperture diaphragm. Almost all of the images present in this atlas have been prepared using this plastic dish method. If one wishes to use an organic solvent such as acetone for fixation, one should use an 80% solution of acetone in H_2O, which will fix and permeabilize cells but not dissolve the plastic dish. Methanol will not dissolve the plastic. Permanox dishes are resistant to these organic solvents but are not yet available in the 35-mm size. This same approach is used to process cells for electron microscopy using ethanol-miscible embedding media and moderate temperature (58°C) polymerization. Thus, one can use plastic dishes directly for most forms of immunocytochemistry.

D.   REAGENTS

The usual procedure we employ for immunofluorescence is the indirect method using a second antiglobulin step conjugated to a fluorochrome. As primary antibodies, we have used polyclonal antibodies obtained from rabbits, rats, mice, goats, and humans (with autoimmune disorders). The appropriate second step would then be an antiglobulin–fluorochrome conjugate made against this species of immunoglobulin (e.g., anti-rabbit IgG). For monoclonal antibodies, which are usually mouse or rat, the second step would be an anti-mouse or anti-rat IgG conjugate. Anti-mouse IgG made in rabbits or goats is commercially available. The usual titer of antiglobulins, and most antibodies in animals, is less than 5% of the total IgG fraction in their sera or plasma. Thus, >95% of the IgG fraction is something else. For a fluorochrome conjugate, >95% of the conjugated antibody in the second step reagent would then be nonspecific. For this reason, there is a great advantage in using affinity-purified fluorochrome conjugates. For example, goat anti-mouse IgG conjugated to rhodamine would be purified by passing it over a mouse IgG affinity column and eluting the high-affinity antibody as a separate fraction. This fraction would then be 100% anti-mouse IgG, and no other rhodamine-conjugated proteins would be present. Such affinity-purified second-step fluorochrome reagents are now available commercially. Jackson ImmunoResearch (Avondale, Pennsylvania) and Cappel Labs (Malvern, Pennsylvania) are examples of two companies that sell such purified reagents. We routinely use only affinity-purified reagents for immunofluorescence.

The primary antibody also contains nonspecific immunoglobulins if it is derived from an animal. Monoclonal antibodies derived from ascites may have other nonspecific contaminating antibodies from the host animal. Antibodies derived from tissue culture supernatants (unless cells are grown in serum-free media) have proteins derived from serum present in the culture medium. Therefore, some purification may be necessary even for monoclonal antibodies to eliminate background. Our usual approach for screening, however, is to test culture supernatants directly (see Appendix). The concentration of antibody in culture supernatants may be only 1–10 μg/ml, which is at the lower end of the range of the antibody concentrations normally used in immunofluorescence experiments. To test media on formaldehyde-fixed cells, we simply add 10 μl of a 10% saponin solution to 1 ml of supernatant (yielding a final concentration of 0.1% saponin). This medium with saponin is then added directly to the fixed cells (0.5–1.0 ml volume) in the plastic dish and incubated at room temperature for 15–30 minutes. This is followed by a series of short washes in PBS containing 0.1% saponin. Note that the culture medium has a large concentration of serum proteins derived from the calf serum present in the medium. Therefore, nonspecific binding due to a low concentration of antibody protein is minimized. In media from hybridomas grown in serum-free conditions, some carrier protein may be necessary to minimize this nonspecific background problem. For a mouse or rat monoclonal, where the second-step antiglobulin conjugate is made in a goat, 1–5 mg/ml of normal goat globulin is the best carrier protein. This is because the one species of IgG that the goat would not have produced antibodies to would be its own goat IgG. This inclusion of carrier "competing" globulin is generally used in all antibody solutions to minimize nonspecific interactions. This is true also for the second-step reagent. For example, the second step would contain affinity-purified goat anti-mouse IgG conjugated to rhodamine along with 0.1% saponin and 4 mg/ml unconjugated normal goat globulin. Such a carrier protein also allows one to harvest the antibody solution back from the dish after the incubation and store it by freezing until the next experiment. Such solutions have been successfully stored frozen for more than 2 years at −20°C. Antibody proteins are often sensitive to freezing and thawing if the concentration of total protein in the solution is low (<0.5 mg/ml). The inclusion of carrier globulin protects the specific antibody from damage due to freezing and thawing. Further, one may want to pass these solutions through a Millipore filter prior to adding them to the dish to eliminate small aggregates that have developed during storage. The inclusion of carrier protein allows this filtration step without significant loss of the specific antibody. We commonly refreeze, filter, and reuse antibody reagents 10–20 times before noting any significant loss in reactivity. This results in a substantial savings over many experiments and guarantees a dependable reagent for multiple tests.

Our choice of fluorochrome is usually rhodamine. The reasons are that rhodamine has a slower bleach rate than fluorescein, there is less autofluorescence in cells and dishes in the rhodamine spectrum, and rhodamine is less light sensitive during processing, so that all the incubations can be performed in normal room light. The proper microscope filters for rhodamine labels are readily available. Fluorescence microscopes for use with rhodamine should have mercury vapor light sources and not xenon bulbs, since there is very little emission from xenon in the green spectrum compared to mercury vapor. Now that rhodamine conjugates are commercially available, there is usually less reason to use fluorescein, except as a second label for double-label experiments.

E. CONTROLS

Immunofluorescence is a very sensitive method, and controls for specificity are essential for proper interpretations. Immunodiffusion (Ouchterlony) plates are much too insensitive for establishing antibody specificity. Fluorescence is at least a thousandfold more sensitive than routine immunodiffusion analysis. It is best to use a sensitive biochemical assay to determine the components in the cell that react with a certain antibody. It is not sufficient to show that an antibody reagent reacts with the purified antigen of interest; one must also show that this reagent does *not* react with other components in the cell. For proteins, this can be accomplished by either Western blot (immunoblot) (3) analysis of the antibody or immunoprecipitation of metabolically labeled cells followed by sodium dodecyl sulfate polyacrylamide gel electrophoresis and autoradiography. No procedure is foolproof. Some proteins may not contain the amino acid, such as methionine, chosen for the metabolic labeling. Other antigens may be insoluble in the homogenization buffers used. Some antigens may be taken up from the medium and not be synthesized by the cells. In immunoblots, some proteins may be denatured into a conformation that is not recognized by the antibody. Some antigens may be glycolipids or other components that are not shown on gels used to resolve proteins. Thus, the careful characterization of an antibody requires very sensitive techniques that can detect nanogram amounts of antigen. Further, a prudent evaluation of the probable chemical characteristics of the potential cross-reactive antigens in the cell is advisable.

Once it is reasonably well established that the antibody reacts only with the antigen in question, the technique of immunofluorescence itself must be controlled. One must establish that the second-step antiglobulin reagent does not react with the cell by itself. Deletion of the first antibody step is the simplest way to control for this possibility. Substitution of an equivalent amount of a "normal" globulin for the first step is commonly used as a control, but one must be careful that this "normal" globulin is really nonreactive with the cell, since many unexpected antibodies are present in "normal" animal sera. A better control with an affinity-purified first antibody is to use a similarly affinity-purified first-step antibody made to an antigen known to be absent from the cell. Affinity-purified anti-horseradish peroxidase (a plant protein not normally present in animal cells) is a good control. The same is true for monoclonal antibodies, in that a proper control would be another monoclonal of the same subtype IgG or IgM that reacts with an antigen absent from the cell.

High-affinity antibodies to single determinants, such as monoclonal antibodies, often have affinity constants in the range of $0.2-1.0$ $\mu g/ml$. One should see reasonable localization, then, at $10-50$ $\mu g/ml$. If the purified antibody requires a higher concentration, possible nonspecific interactions should be carefully evaluated. For whole sera, where the total IgG concentration may be $10-20$ mg/ml, the specific antibody may represent only 1% of the total IgG. Thus, a $1:20$ dilution would yield a specific antibody concentration of $5-10$ $\mu g/ml$. Since this specific antibody may be directed against multiple sites on a protein, the concentration of antibody against a single antigenic determinant may be $1-2$ $\mu g/ml$. Thus, characterizing antibodies as to their potential titer of dilution may reflect only the titer in the serum and not relate to the inherent specificity of the antibody present in the serum. Even so, the high dilution of the serum will progressively lower the concentration of other contaminating antibodies. For some small peptides that are very antigenic, the titer of the IgG fraction may be 50% specific. This implies that at a $1:10,000$ dilution with 10 mg/ml total IgG, the specific antibody concentration will be 0.5 $\mu g/ml$, within the association constant range for many antibodies. The nonspecific globulin concentration would also be 0.5 $\mu g/ml$ for a broad mixture of other antibodies, and for any one antibody would be well below the association constant for binding. Unfortunately, the titer of antisera to most antigens never approaches 50%, but is usually $0.1-5\%$ of the total IgG. Thus, affinity purification of the primary antibody, if possible, is a great help in producing a clear result. To perform affinity purification usually requires milligram quantities of the antigen, which are not always possible to obtain. Therefore, immunofluorescence interpretations are usually limited by the purity of the reagents.

F.   PHOTOGRAPHY

The direct image of fluorescence samples in the microscope is colorful. Because of barrier filters, a rhodamine image is red and a fluorescein image is green. When substantial background autofluorescence is present, the specific signal may be one color and the autofluorescence another. This makes photography using color film an obvious choice. However, for cultured cells the autofluorescence background is almost nonexistent. In these samples, anything one can see is the same color, and color film photography is not particularly necessary. In fact, color film can produce problems for immunofluorescence images. When color film overexposes, the recorded image is usually yellow. Thus, in a rhodamine image where all the light being recorded is automatically red due to barrier filters, overexposed regions appear yellow on this red background. This creates a false impression of specificity that is not present in the sample. Color films are generally less sensitive than black and white films, requiring longer exposure times or in some cases being unable to record the image due to photobleaching of the fluorescence before an image is recorded. Slides made from color transparency film are very dark and are often not useful for projection. Therefore, we usually use only black and white film for recording fluorescence images. High-speed, low-grain film such as Ilford XP-1 is available with ASA ratings of 1600. This film requires special processing, which makes it less convenient for routine use. Kodak Tri-X film has an ASA rating of 400 in normal developers, but a 1600 ASA rating when developed using a two-step developer, such as Diafine. This developer also has the advantage that the time of development is not critical. Each of the two development steps can be left for the minimum time of 3 minutes or for 1 hour without affecting the results. This makes rapid development with flexible times quite convenient. This film used in a 35-mm camera back on a fluorescence microscope is the most convenient system we have found. Polaroid films are convenient but have significantly less contrast and clarity. The resultant prints are much less flexible in comparison to the variations in contrast and exposure one can achieve with normal negative film. Thus, it is important for a fluorescence microscope facility to have an ajoining darkroom with enlarger and rapid print processors to make the images available quickly.

For some images, direct recording on photographic film is not practical because of the low intensity of fluorescence emission. This is a case for the application of image intensification technology, which we have recently reviewed in detail (6). It should be pointed out that image intensification will not increase the signal-to-noise ratio of images but will only increase the level of detectability of weak signals. If the image can be seen using low-magnification eyepieces in a darkened room, then the image can be recorded using intensifier systems. In most cases, one should not expect the intensifier systems to be able to produce the same resolution possible with photographic film.

One of the main limitations in recording immunofluorescence images is photobleaching of the fluorochrome. The usual fluorescence sample using a high numerical aperture objective will be bleached in 0.2–0.5 minutes for fluorescein, or 2–3 minutes for rhodamine. If an image is not recorded on film in this time, then it cannot be recorded. However, there are materials which can be added to the glycerol mounting medium that can dramatically extend the lifetime of these fluorochromes. We have successfully used both (5%) N-propyl gallate (1) and (1 mg/ml) p-phenylenediamine (free base) (2) for this purpose, and these are routinely included in mounting fluorescein-labeled samples for photography. They extend the lifetimes of both rhodamine and fluorescein considerably and allow photography not possible without them. Their effect on the primary fluorescence intensity is minimal. N-Propyl gallate in glycerol tends to crystallize with time and p-phenylenediamine tends to get darker with time. Storage of p-phenylenediamine in a freezer in the dark is recommended. It should be noted that the free base of p-phenylenediamine is the proper reagent, *not* the dihydrochloride derivative.

APPENDIX

Protocol for Routine Immunofluorescence of Cultured Cells Using Hybridoma Media

1. Plate cells in 35-mm tissue culture plastic dishes on the previous day
2. Wash off culture media using PBS (with Ca^{2+}, Mg^{2+}), $\times$ 2
3. Fix cells by adding 3.7% formaldehyde in PBS for 10 minutes at 23°C
4. Wash off fixative solution using PBS, $\times$ 5
5. Incubate cells on a tilting platform in primary mouse antibody as follows:
   To 1 ml mouse hybridoma supernatant medium (undiluted) add 10 μl 10% saponin
   Incubate for 30 minutes at 20–23°C
   Harvest and freeze antibody solution for reuse
6. Wash in PBS containing 0.1% saponin, $\times$ 5, 10 minutes total
7. Incubate in second step:
   Affinity-purified goat anti-mouse IgG conjugated to tetramethylrhodamine (from Jackson ImmunoResearch), 50 μg/ml in PBS, 0.1% saponin, 4 mg/ml normal goat globulin, for 15 minutes at 23°C
   Harvest and freeze antibody solution for reuse
8. Wash in PBS containing 0.1% saponin, $\times$ 5, 10 minutes total
9. Wash in PBS, $\times$ 2, 5 minutes total
10. Mount in buffered glycerol under a #1 25-mm diameter circular coverslip
11. Examine using a 63$\times$, oil (N.A. 1.4) planapochromat objective and rhodamine epifluorescence optics (Zeiss RA standard upright microscope)

Note: For antigens present in mitochondria or nuclei, 0.1% Triton X-100 treatment (3 minutes at 20–23°C, in PBS) is recommended immediately following formaldehyde fixation.

REFERENCES

1. Giloh, H., and Sedat, M. (1982). Fluorescence microscopy: Reduced photobleaching of rhodamine and fluorescein protein conjugates by N-propyl gallate. *Science* **217**, 1252–1255.
2. Platt, J. L., and Michael, A. F. (1983). Retardation of fading and enhancement of intensity of immunofluorescence by p-phenylenediamine. *J. Histochem. Cytochem.* **31**, 840–842.
3. Towbin, H., Staehelin, T., and Gordon, J. (1979). Electrophoretic transfer of proteins from polyacrylamide gels to nitrocellulose sheets: Procedure and some applications. *Proc. Natl. Acad. Sci. U.S.A.* **76**, 4350–4354.
4. Weber, K., Rathke, P. C., and Osborn, M. (1978). Cytoplasmic microtubular images in glutaraldehyde-fixed tissue culture cells by electron microscopy and by immunofluorescence microscopy. *Proc. Natl. Acad. Sci. U.S.A.* **75**, 1820–1824.
5. Willingham, M. C. (1983). An alternative fixation-processing method for pre-embedding ultrastructural immunocytochemistry of cytoplasmic antigens: The GBS procedure. *J. Histochem. Cytochem.* **31**, 791–798.
6. Willingham, M. C., and Pastan, I. H. (1983). Image intensification techniques for detection of proteins in cultured cells. *In* "Methods in Enzymology" (S. Fleischer and B. Fleischer, eds.), Vol. 98, pp. 266–267. Academic Press, New York.

2

BRIGHT-FIELD IMAGES OF FIXED CELLS

—————— PLATE 1 ——————

Phase-Contrast Image of a Flattened Cultured Mouse Fibroblast. This is a Swiss 3T3 cell grown on a plastic substratum, fixed in glutaraldehyde, and photographed using phase-contrast optics. The outline of the flattened cell in the center is shown by the arrowheads. The nucleus (N) also contains darker nucleoli which are just beyond the plane of focus. The cytoplasm contains reticular structures that are difficult to resolve, but some of the wormlike elements in the more peripheral cytoplasm are mitochondria. Other cells nearby are less spread out. (Mag., ×1220.)

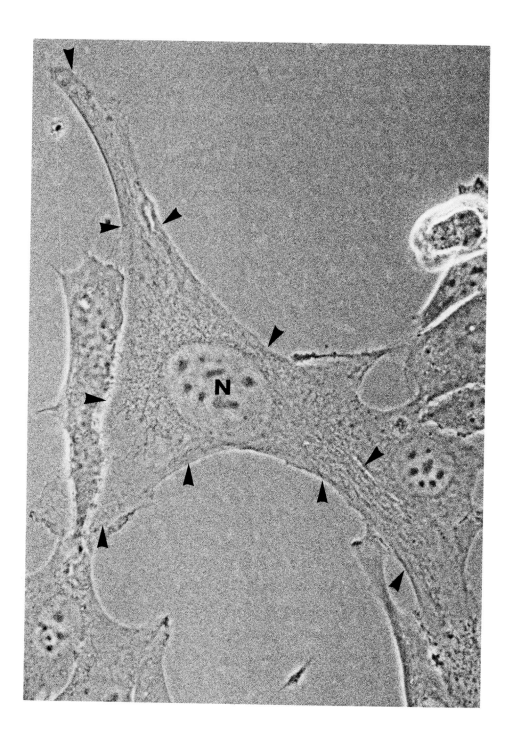

————————— PLATE 2 —————————

Phase-Contrast Images of Ruffles and Macropinosomes in Flattened Cells. A collection of spreading Swiss 3T3 cells is shown here using phase-contrast optics. The nuclei in this group of cells are labeled (N). At the peripheral edges of many cells there are ruffles (large arrows). The folding of these ruffles back onto the adjacent cytoplasm produces phase-lucent macropinosomes in the cytoplasm, some of which are labeled with arrowheads. These are the structures identified by Lewis in 1931 (26) that mediated "pinocytosis." The cell in the lower part of this plate shows blebs on its surface (small arrows), which appear phase-dark. (Mag., ×1220.)

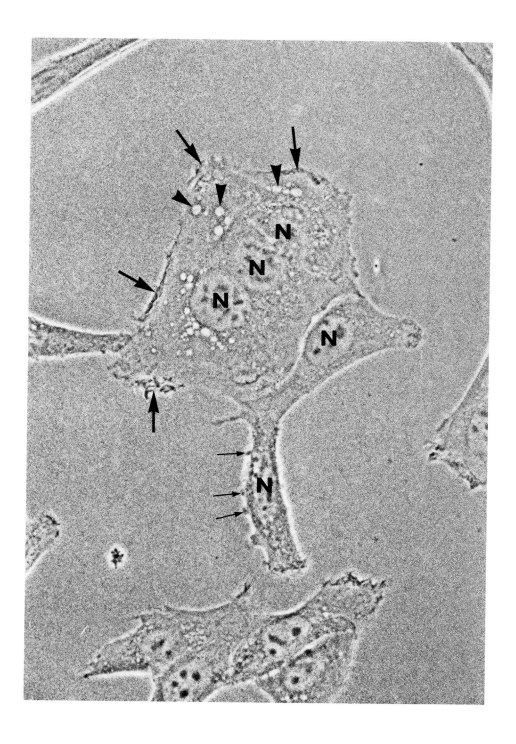

PLATE 3

Phase-Contrast Image of Migrating and Mitotic Cells. This plate shows a group of Swiss 3T3 cells during active migration. The cells are not well spread and show considerable leading lamellar ruffle activity (large arrows). These ruffling lamellae generally precede the advancing cell margin and also produce large amounts of pinosomal entry as shown by the accumulation of phase-lucent macropinosomes in the cytoplasm of these cells. In the lower panel, two daughter cells after mitosis show the characteristic surface blebs (small arrows) seen in telophase with the contraction of the cleavage furrow. (Mag., ×1220.)

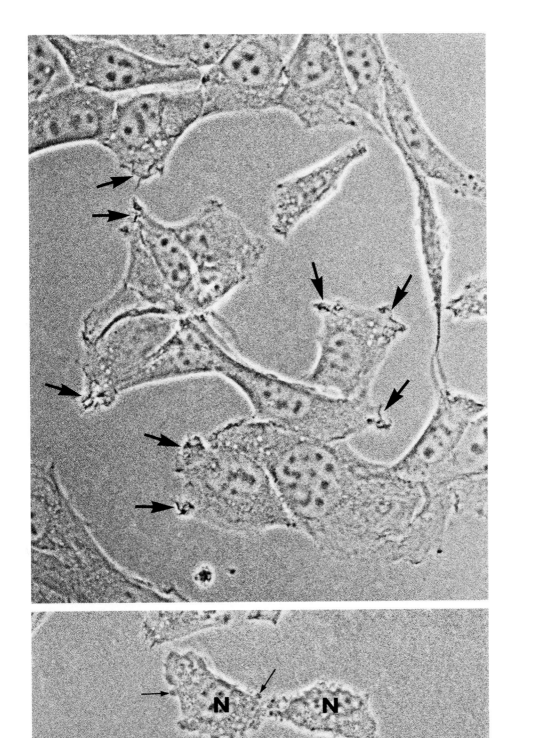

——————— PLATE 4 ———————

Bright-Field Image of a Cultured Fibroblast Stained with OTO. This image shows the enhanced contrast and preservation of lipid droplets (arrows) seen in a Swiss 3T3 cultured mouse fibroblast using the osmium–thiocarbohydrazide–osmium (OTO) technique (56). This method deposits dark osmium in lipid-containing structures, which otherwise are seen only as phase-refractile droplets in living cells. This image demonstrates a type of major cellular inclusion (lipid droplet) that is not visualized in any of the immunofluorescence images, except by exclusion of diffusely distributed cytoplasmic markers. (Mag., ×1250.)

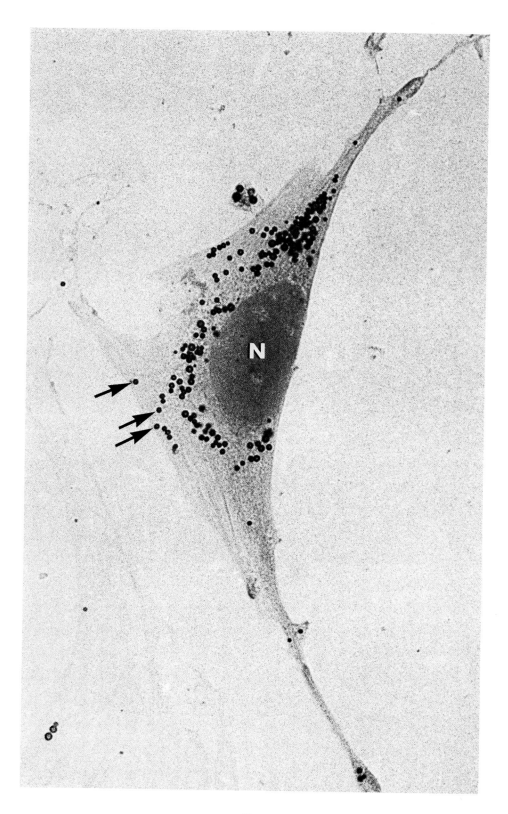

3

IMMUNOFLUORESCENCE IMAGES

─────────── PLATE 5 ───────────

Homogeneous Labeling of the Cell Surface in Epithelioid Cells. This image shows a dense culture of KB cells, an epithelioid human carcinoma cell line, labeled using an antibody to the extracellular domain of the epidermal growth factor receptor (EGF-R1) (4). In the absence of added EGF, the receptor remains diffusely distributed on the cell surface. The image over the center of the cells is weak and diffuse, but when a larger amount of the surface label can be seen in a vertical plane at the edges of the cells, the accumulated image is brighter. This produces the bright lines of fluorescence at the cell margins, even though the concentration of label per unit area of membrane area at the cell margins is the same as that over the center of the cells. (Mag., ×1350; formaldehyde fixed prior to antibody incubation, no permeabilization.)

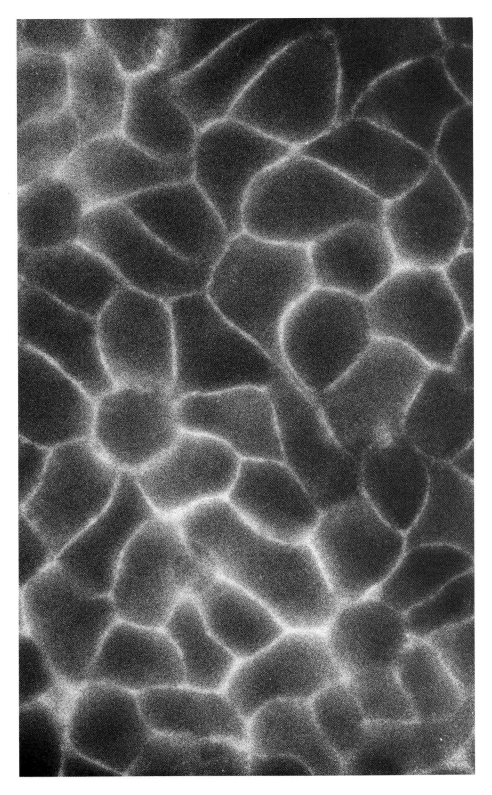

——————— PLATE 6 ———————

Homogeneous Labeling of the Cell Surface in Flat Epithelioid Cells. This image is similar to that shown in Plate 5, except that the KB cells are flatter. The cell margins are less distinct, and the surface pattern appears weaker and more diffuse, with less of the bright marginal accumulation seen in the more rounded cuboidal cells (Plate 5). This pattern is produced in cells that have approximately 100,000 receptor sites per cell on the surface. Where the cell margin is also flattened (arrow), the image appears weak and less well defined than the image at the vertical margins seen in tightly packed cells as shown in Plate 5. (Mag., ×1350.)

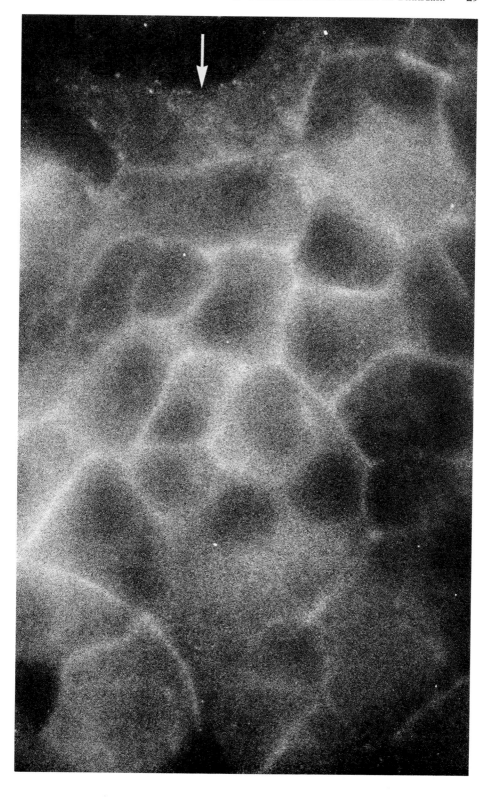

—————— PLATE 7 ——————

Heterogeneous Labeling of the Cell Surface on Adjacent Cells. This image demonstrates labeling of A431 cells using a monoclonal antibody (MC 101) (13, 37) to a surface carbohydrate determinant that is strongly expressed, weakly expressed, or not at all expressed by different cells in the same culture. The cells denoted as (A) are heavily labeled in a diffuse surface pattern that shows labeled retraction fibers (structurally similar to microvilli) at the cell edge (arrow) and an irregular surface. Those denoted as (B) are unlabeled, and those denoted as (C) are weakly labeled. Such heterogeneity is sometimes found with carbohydrate determinants on the cell surface. (Mag., ×1350.)

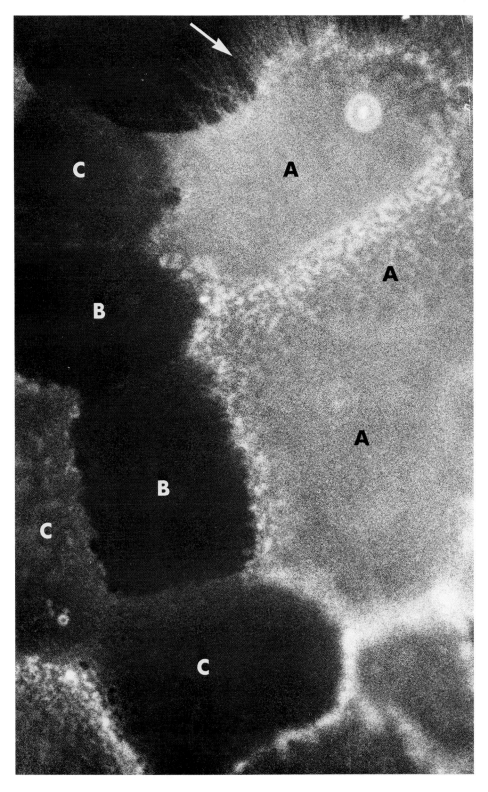

———————— PLATE 8 ————————

Double-Label Images of Heterogeneous and Homogeneous Surface Patterns. These images are from a double-label experiment using in (A) an antibody against a carbohydrate determinant that is heterogeneously expressed on these A431 cells (MC 101) (13, 37) with rhodamine labeling. In (B) is the fluorescein image of the same cells using an antibody (MC EGF-R1) (4) to the EGF receptor that is homogeneously distributed on the cell surface. Note the cells denoted [1] have very little labeling with rhodamine (MC 101), but the same cells show uniform diffuse surface labeling with fluorescein (MC EGF-R1). On the other hand, cells denoted [2] have intense labeling with rhodamine and a uniform diffuse pattern with fluorescein. (Mag., ×1420.)

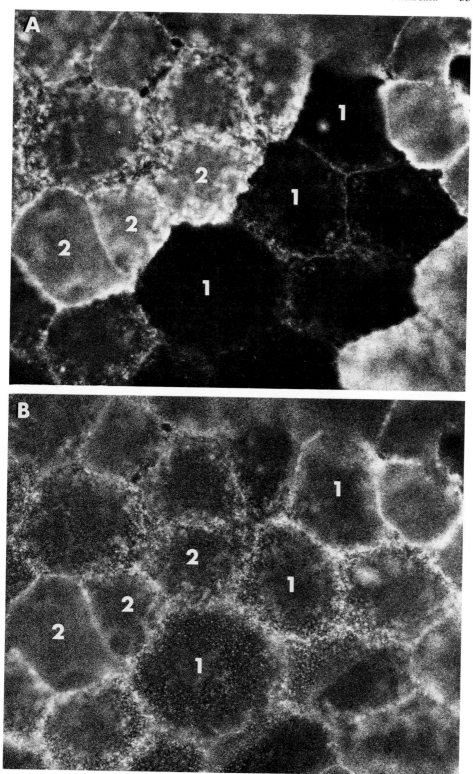

———————— PLATE 9 ————————

Homogeneous Labeling on the Inner Surface of the Plasma Membrane. This image shows A431 cells labeled using a monoclonal antibody against the p21 Harvey murine sarcoma virus (MSV) v-*ras* protein (MC YA6-172) (49, 55). This protein is present in the inner surface of the plasma membrane in these human carcinoma cells. The pattern shown is similar to that seen for a diffusely distributed label on the outer surface of the plasma membrane, except that this labeling is not seen without a permeabilization step; in this case Triton X-100 treatment was used after formaldehyde fixation. (Mag., ×1250.)

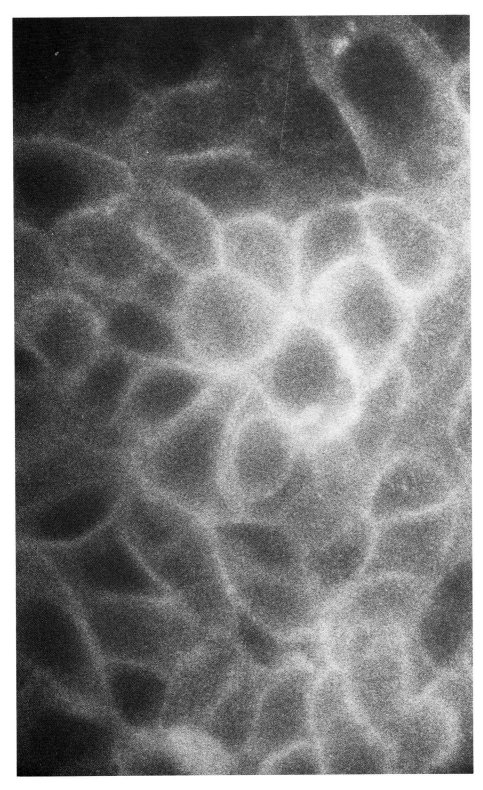

———————— PLATE 10 ————————

Labeling of the Inner Surface of the Plasma Membrane. This image shows labeling of the inner surface of the plasma membrane in a human bladder carcinoma cell line (RT4) using a monoclonal antibody (MC YA6-172) (49, 55) against Harvey MSV v-*ras*. These cells are more rounded and variable in shape than the A431 cells shown in Plate 9, but show the same bright diffuse surface-related pattern after fixation and permeabilization. (Mag., ×1150.)

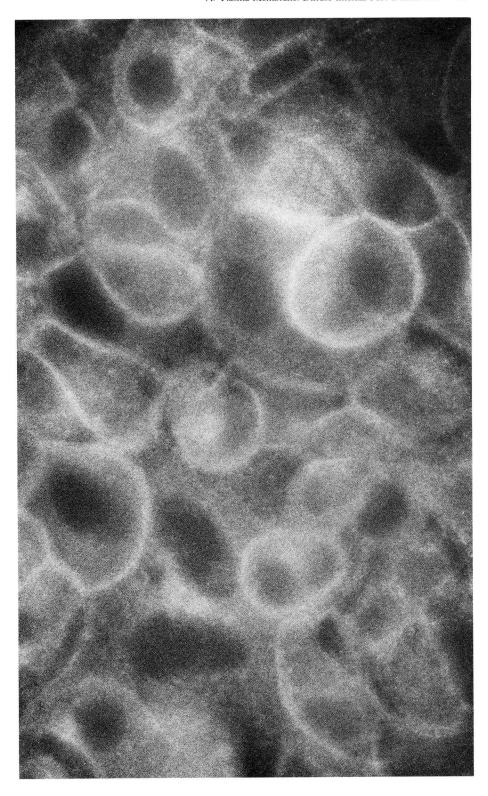

—————— PLATE 11 ——————

Labeling of the Inner Surface of the Plasma Membrane in Flat Cells. Harvey MSV-transformed NIH3T3 cells show a diffuse pattern of v-*ras* on the inner surface of the plasma membrane using (MC YA6-172) (49, 55). In this image, the diffuse pattern can be seen as a weak labeling of the entire cell outline. These cells also show more concentrated labeling in the perinuclear Golgi region; this pattern may represent recycling plasma membrane derived from the surface or overproduction of this protein trapped in Golgi-associated elements (arrows). Note that surface ruffles also show bright labeling at the cell periphery (arrowheads) because of the increased amount of label projected over this area by the vertical ruffle structures. Note that the diffuse surface label also is present over the nuclear regions (N), unlike the sparing of these regions that would appear with the label located diffusely in the cytoplasm. (Mag., ×1150.)

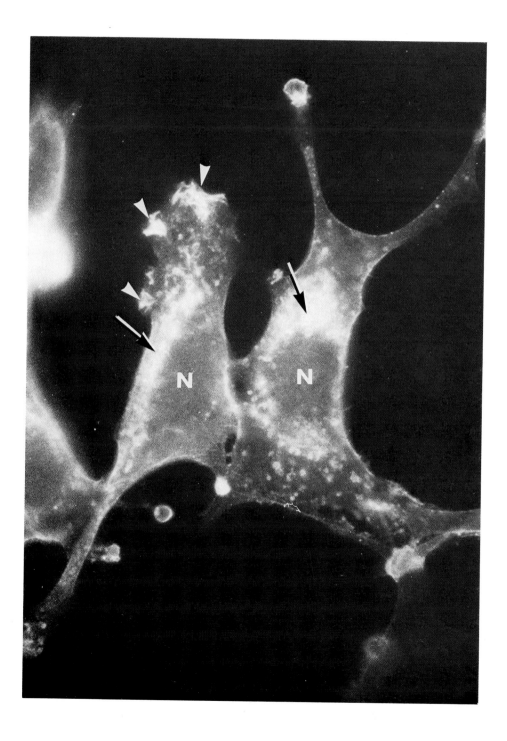

———————— PLATE 12 ————————

Labeling of the Inner Surface of the Plasma Membrane in Spindly Cells. This image shows Harvey
MSV-transformed NIH3T3 cells that are spindly and less flat than those shown in Plate 11.
Here the cell margins are more rounded, and more of the cell surface is projected into a vertical
image, resulting in a greater accumulation of label fluorescence in that region. The resulting
image shows a bright linear margin along the cell borders that represents a plasma membrane-
related pattern. This is only seen for this antigen in cells that have been permeabilized using, in
this case, Triton X-100 after formaldehyde fixation. Some perinuclear-increased fluorescence is
also evident in these cells, but note that the regions directly over the nucleus show a diffuse
bright image, indicating that some of this labeling is surface-related. (Mag., ×1150.)

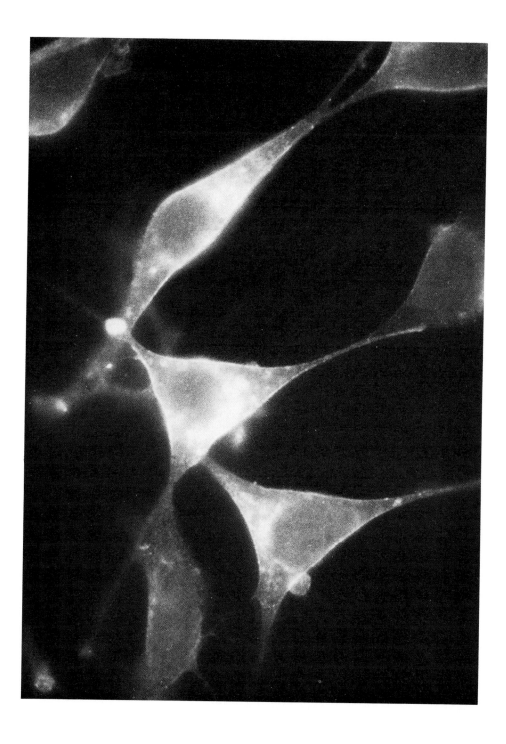

———————   PLATE 13   ———————

Labeling of the Inner Surface of the Plasma Membrane in Rounded Cells. (A) shows an elongated Harvey MSV NIH3T3 fibroblastic cell that is so spindly that its edges are highly rounded. The very bright linear pattern of label (YA6-172) (49, 55) on the inner surface of the plasma membrane is evident. In (B), daughter cells of a recent mitosis are shown which demonstrate this bright submembranous pattern. (Mag., ×1150.)

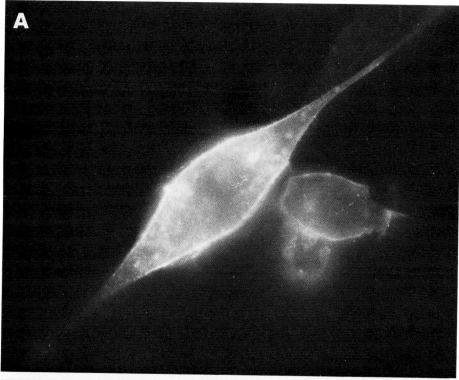

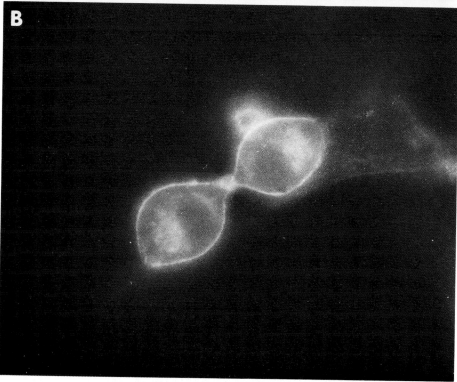

——————— PLATE 14 ———————

Clathrin-Coated Pits in a Flattened Cell. This shows a high magnification image of a portion of a single flattened Swiss 3T3 cells using anti-clathrin antibody (21, 50). The small punctate dots over the entire cell image represent randomly distributed clathrin-coated pits at the plasma membrane. In the perinuclear region, a more confluent fluorescence is due to the small coated structures associated with the Golgi system. (Mag., ×2000.)

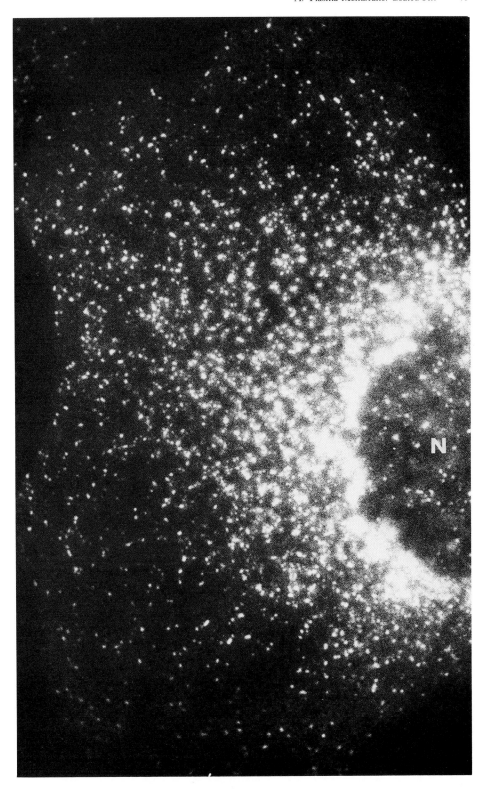

——————— PLATE 15 ———————

Low-Magnification View of Clathrin-Coated Pits. This image shows a low magnification view of Swiss 3T3 cells using anti-clathrin antibody (21, 50). Note the randomly distributed dots over the cell surface that represent plasma membrane coated pits on both the upper and lower cell surfaces. Also, the accumulation of clathrin-coated structures in the Golgi system can be seen in the perinuclear region of these cells. (Mag., ×1250.)

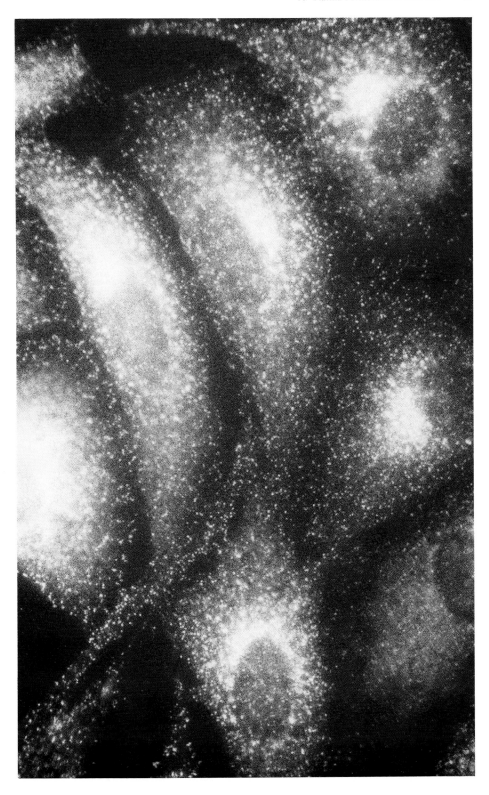

--------- PLATE 16 ---------

Clathrin-Coated Pits in Spindly Chick Fibroblastic Cells. This image shows the pattern of clathrin localization (21, 50) in spindly chick embryo fibroblasts. The plasma membrane-coated pits appear as bright dots over the entire cell image, and the clathrin-coated structures of the Golgi system show a perinuclear eccentric accumulation. (Mag., ×1350; N, nucleus.)

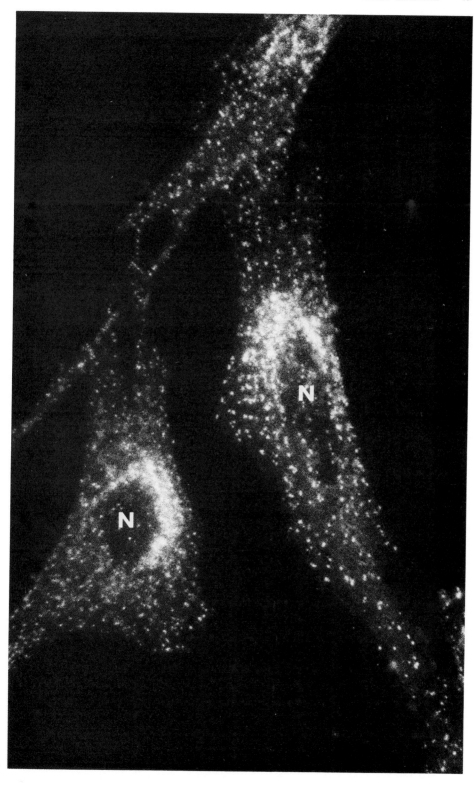

——————— PLATE 17 ———————

Elongated Chick Fibroblasts with Anti-Clathrin. These panels show examples of both the finely punctate distribution of plasma membrane-coated pits in elongated chick fibroblasts and the stretched-out pattern of the coated structures of the Golgi system (21, 50). Some of these Golgi elements can be seen to be oriented parallel to the long axis of the cell with a reticular, interrupted distribution. (Mag., ×1350, N, nucleus.)

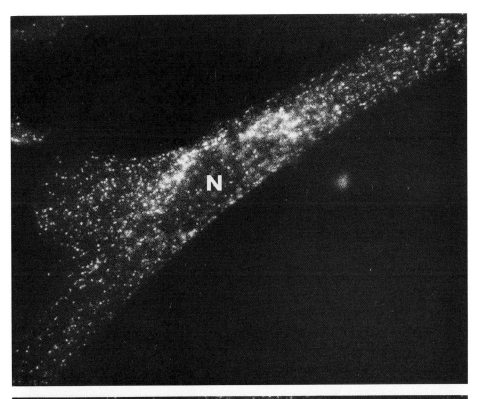

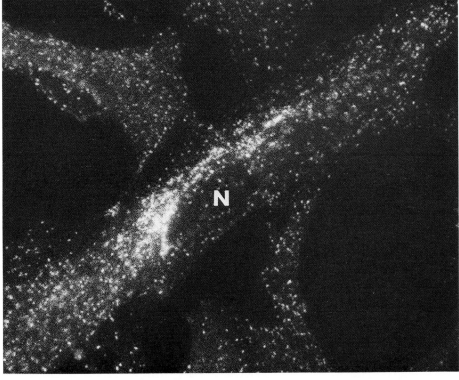

————— PLATE 18 —————

Leading Lamellar Ruffles in a Migrating Cell. This image shows a chick embryo fibroblast fixed during migration toward the top of the panel and labeled using anti-actin antibody (24, 58). At the leading lamellae, the bright linear fluorescence (arrows) represents a vertically folded lamellar ruffle, projecting upward from the substratum. These ruffles contain a diffuse actin microfilamentous network which labels strongly with anti-actin, in addition to the brighter image created by the vertically folded nature of the ruffle. (Mag., ×1250.)

——————— PLATE 19 ———————

Lamellar Ruffling in a Spread Cell. This image shows a chick embryo fibroblast flattened on its substratum with lamellar ruffles at two ends of the cell (arrows) labeled using anti-actin antibody (24, 58). (Mag., ×1250.)

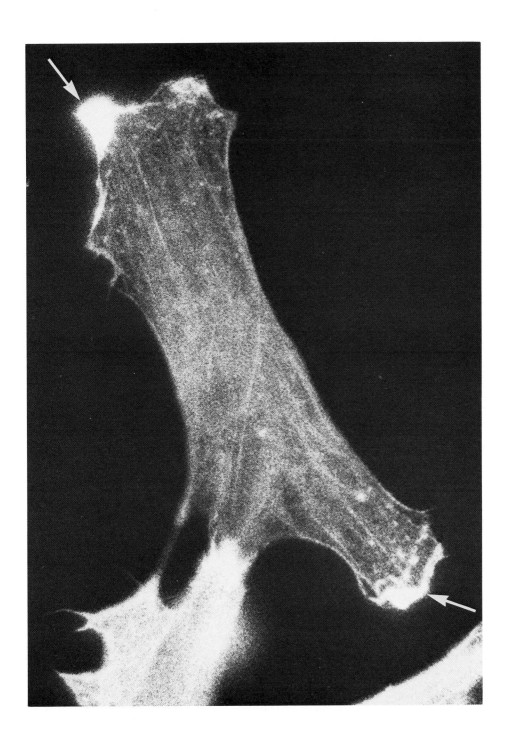

———————— **PLATE 20** ————————

Postmitotic Spreading Cells with Active Lamellar Ruffling. This image shows postmitotic daughter cells (Swiss 3T3) labeled using anti-actin antibody (24, 58). The initial spreading reaction from rounded mitotic cells is accompanied by intense lamellar ruffling activity. These ruffles label well with anti-actin (arrows) and encompass a large portion of the cell lamellar margin. (Mag., ×1250; N, nucleus.)

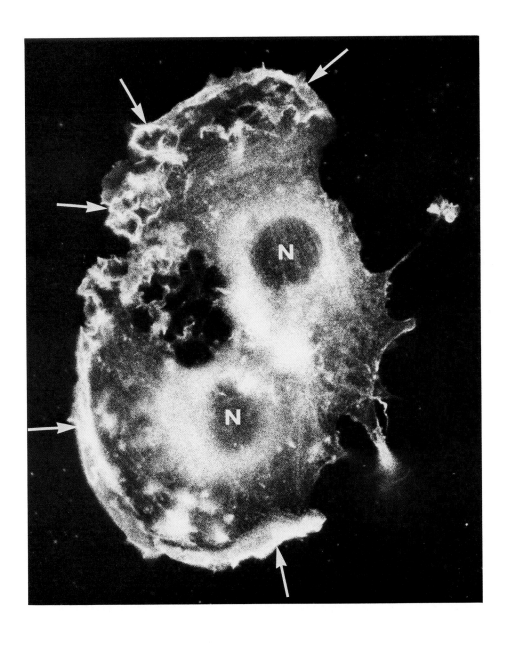

—————— PLATE 21 ——————

Ruffles Demonstrated Using Anti-Alpha Actinin. This image shows chick embryo fibroblasts labeled using anti-alpha actinin (23, 43). This demonstrates that lamellar ruffles (arrows) also label well with this antibody, in addition to labeling with anti-actin. A characteristic feature of anti-actinin is the periodic nature of its localization in other intracellular sites, particularly along microfilament bundles. (Mag., ×1150.)

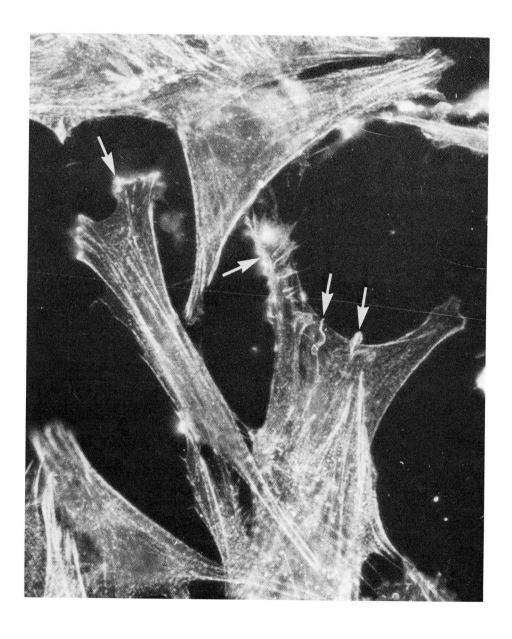

——————— PLATE 22 ———————

Lamellar Ruffles Labeled by a Diffuse Cytoplasmic Marker. This image is of a Swiss 3T3 cell labeled using anti-calmodulin antibody (57). Calmodulin is distributed in a diffuse cytoplasmic distribution, but even though it is not especially concentrated in ruffles, this diffuse pattern shows the enhancement of the ruffle image (arrows) due to its vertically reinforced image. (Mag., ×1450; N, nucleus.)

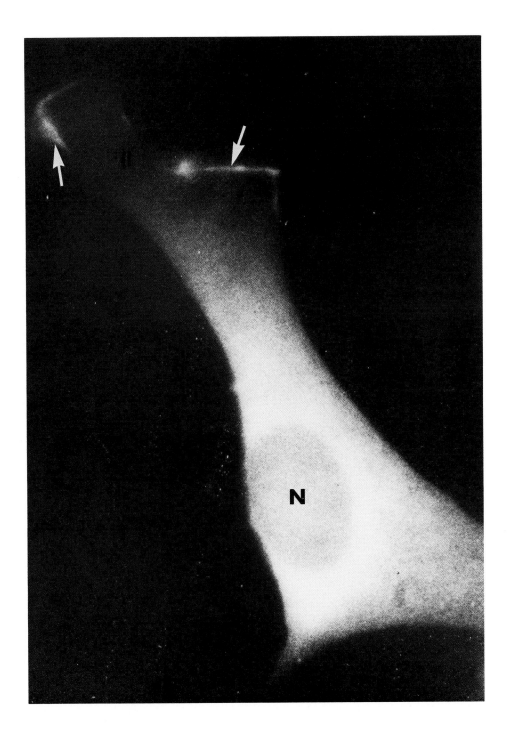

—————— PLATE 23 ——————

Zeiotic Blebs at the Plasma Membrane. Blebs (arrows) are protrusions from the plasma membrane that occur under certain conditions of cytoskeletal damage or increased intracellular pressure. They can be labeled by antibodies to any component of the soluble cytoplasm that is not immobilized to other structures. The image in (A) shows blebs labeled using anti-tubulin in a Swiss 3T3 cell treated with vinblastine (100 μM, 1 hour) (38). The image in (B) is of mitotic cells in telophase that are undergoing contraction of the cleavage furrow, which increases intracellular hydrostatic pressure. These are labeled using antibody to calmodulin, a diffusely distributed cytoplasmic marker. (Mags.: A, $\times 1350$; B, $\times 1000$.)

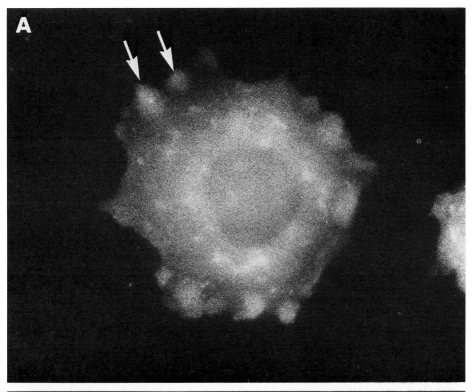

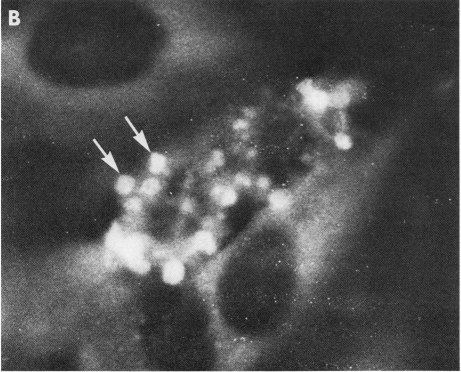

——————— PLATE 24 ———————

Focal Adhesions to the Substratum in Flattened Cells. This image shows the pattern of focal adhesions (focal contacts) to the substratum in chick embryo fibroblasts as detected using an antibody to vinculin (15) (arrows). Note the interrupted, elongated nature of these contacts. The plane of focus for these structures is very narrow, lying just below the cell at the substrate level. (Mag., ×1250.)

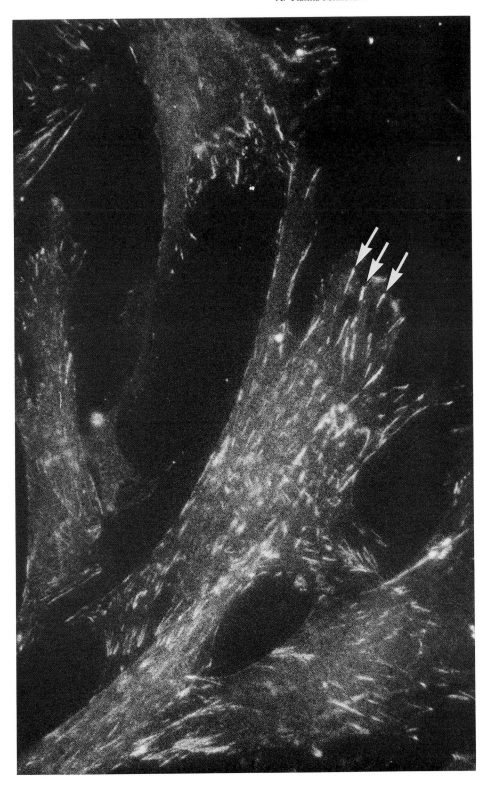

———————— PLATE 25 ————————

Focal Adhesions to the Substratum in Rounded Cells. This image shows the unusual focal contacts present in SR-NRK cells, a rounded, transformed cell type. The focal contacts are visualized using anti-actin antibody, rather than anti-vinculin. This demonstrates the very high concentration of actin in these regions. These structures lie in a very narrow focal plane just under the cell. (Mag., ×1350.)

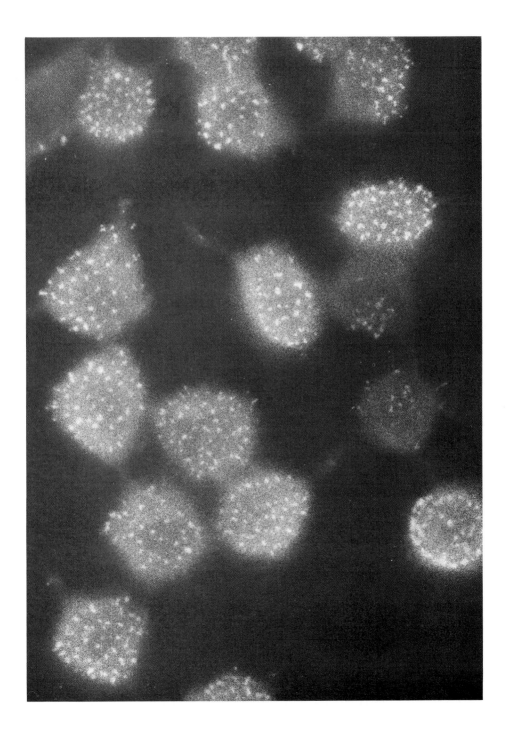

———————— PLATE 26 ————————

Fibronectin Distribution in Flattened Primary Cells. This image shows the pattern of fibronectin in and on flattened primary chick embryo cells (antibody a gift from Dr. K. Hedman) (19). In this dense region, the majority of the fibronectin can be seen as an extracellular matrix component, covering the cells in a web of fibrils. Some of the intracellular fibronectin can barely be seen in the underlying cells. (Mag., ×1250.)

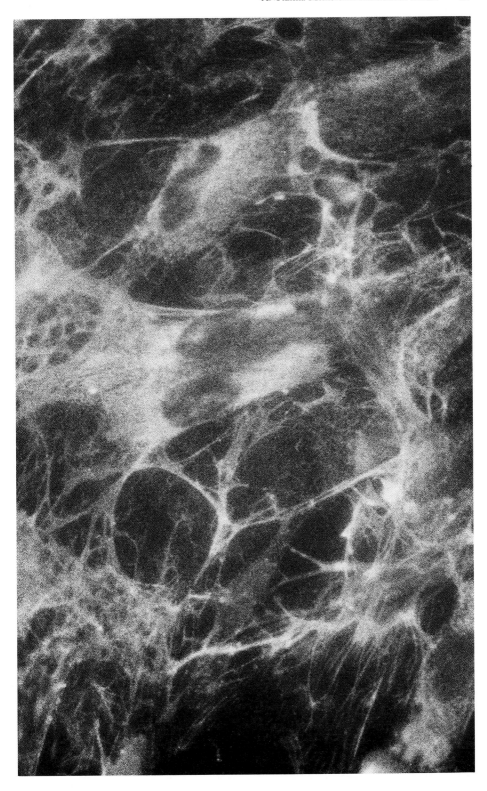

——————— PLATE 27 ———————

Fibronectin in Intracellular and Extracellular Patterns. This image shows a low-density area of a chick embryo fibroblast culture labeled with anti-fibronectin (a gift from Dr. K. Hedman). There are some bright extracellular fibrils (arrowhead) attached to the cell surface, but also note the intracytoplasmic distribution, which spares the nuclear image (N). This pattern is characteristic of the endoplasmic reticulum, a major site of accumulation of fibronectin prior to its secretion. (Mag., ×1250.)

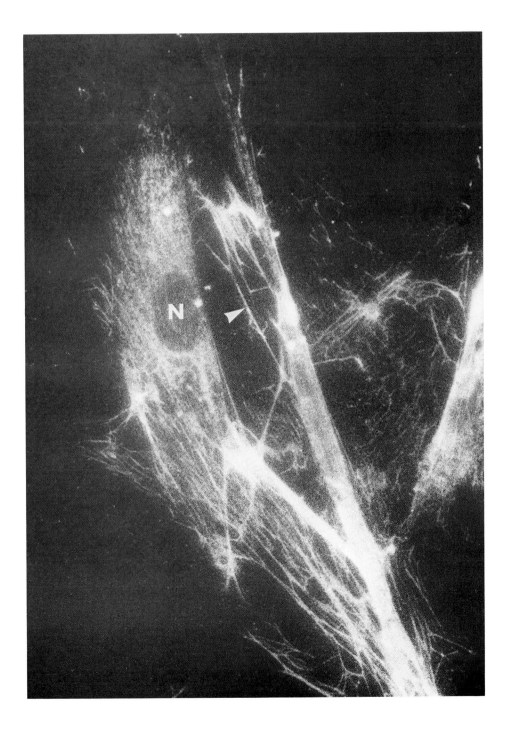

——————— PLATE 28 ———————

Diffusely Distributed Extramembranous Cytoplasmic Antigen. This image shows Swiss 3T3 cells labeled using anti-calmodulin (57), a diffusely distributed protein present in the extramembranous cytoplasm. The nuclei (N) appear unlabeled (nuclear sparing) and there is increased brightness in the perinuclear region due to the increased thickness of the cell near the nucleus. There is not, however, any actual increase in antigen concentration in this location. Other smaller organelles, such as lysosomes, also exclude the label, but this is more difficult to resolve. The edge of the cell is so thin that the brightness decreases significantly at the cell margin, except for folded ruffles, which reinforce the brightness due to their increased thickness (arrow). (Mag., ×1350.)

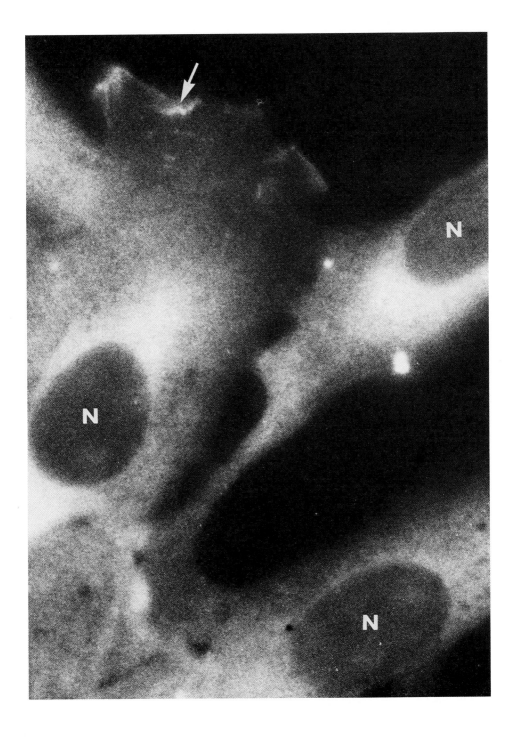

——————— PLATE 29 ———————

Diffuse Extramembranous Cytoplasm. This image shows an NIH3T3 cell labeled using a monoclonal antibody (F1/C5) that detects a protein diffusely distributed in the extramembranous
cytoplasm (a gift from Dr. J. T. August). Note the nuclear sparing and the sparing of smaller
organelles (such as lysosomes) in the perinuclear area. (Mag., ×1350.)

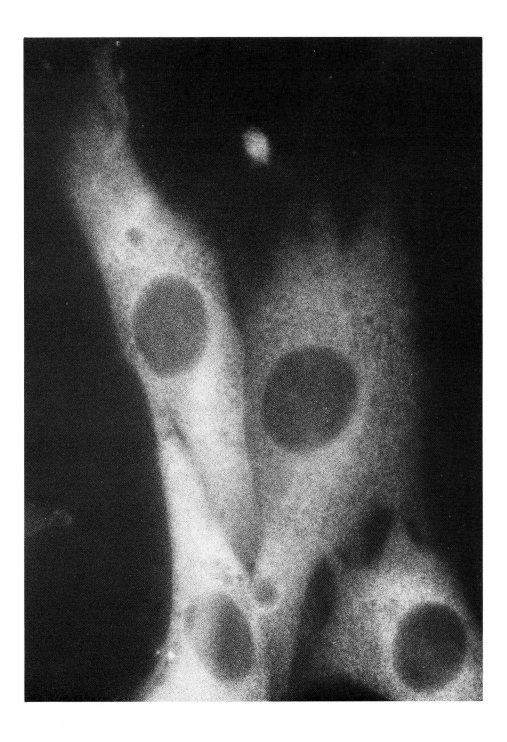

———————————— PLATE 30 ————————————

Cellular Labeling Produced by a Transfected Gene. This image shows the localization of CAT, a bacterial protein synthesized by these CV-1 monkey cells following transfection of the CAT gene (17). Using antibodies to CAT, the protein can be seen to be expressed in two cells in this field and to be distributed both in the extramembranous cytoplasm and within the nucleus. [A, rhodamine anti-CAT fluorescence; B, phase contrast image of the same field (Mag., ×1050).]

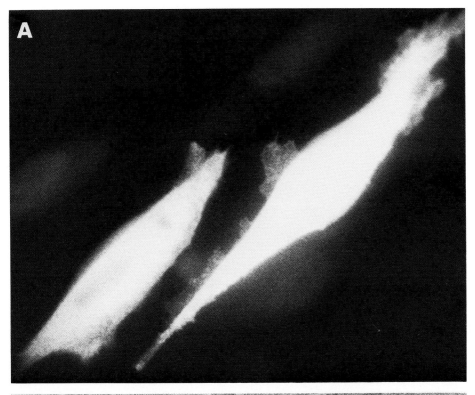

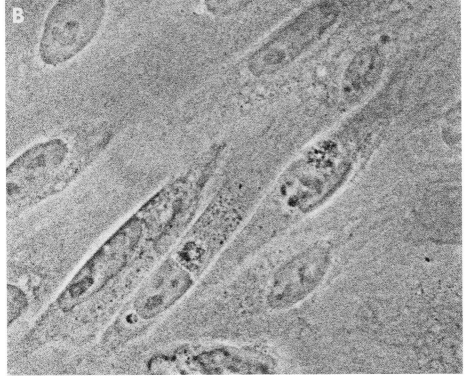

PLATE 31

Focal Aggregates in the Cytoplasm of a Viral Protein. This image shows a chick embryo cell infected with SR-avian sarcoma virus and labeled using a monoclonal antibody to p27, a viral structural protein (a gift from Dr. N. D. Richert) (35). The virus particles assemble at the plasma membrane, incorporating this protein at high concentration, and the bright dots shown in this image may represent individual viral particles during their assembly. (Mag., ×1250.)

———————   PLATE 32   ———————

Mitochondrial Antigen. This image shows the display of mitochondria in Swiss 3T3 cells using an antibody against F_1-ATPase, a mitochondrial marker (a gift from Dr. E. Racker) (11, 34). Note the elongated wormlike shape of these organelles and their occasional branching patterns visible throughout the entire cytoplasm. (Mag., ×1350.)

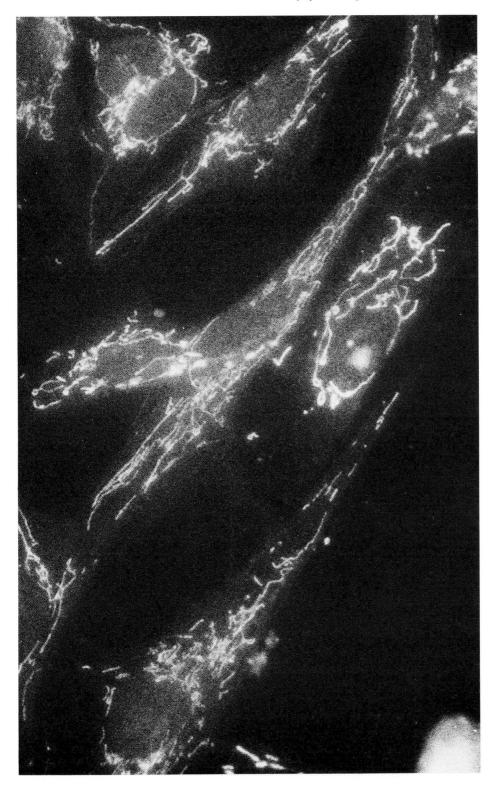

PLATE 33

Mitochondrial Antigen. This image at high magnification shows the extension of mitochondria from the perinuclear region and extending out cell processes. This pattern is characteristic of mitochondria in flattened cultured cells; the antibody used was anti-F_1-ATPase (a gift from Dr. E. Racker) (11, 34). (Mag., $\times 1580$.)

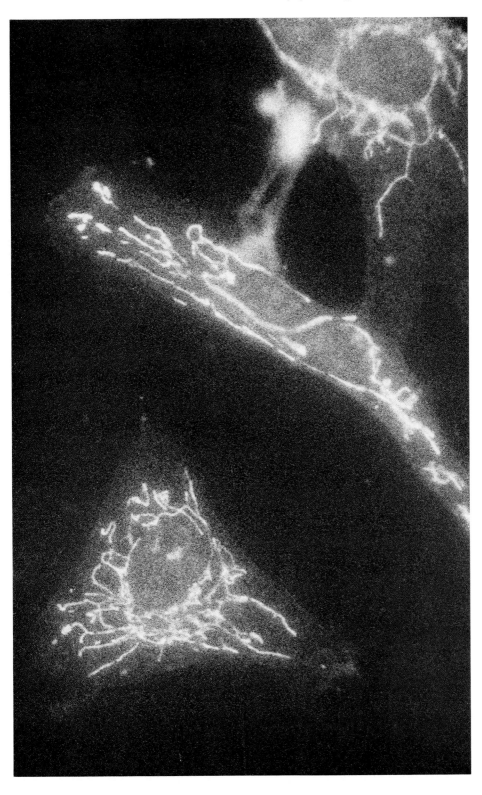

—————— PLATE 34 ——————

Autofluorescence Image of Mitochondria. This image was generated by fixing Swiss 3T3 cells in 1% glutaraldehyde, followed by treatment with 0.1 M glycine (pH 10) (48). The cells show an intense autofluorescence selectively in mitochondria when viewed using rhodamine epifluorescence optics. No antibody or rhodamine was used. (Mag., ×1150.)

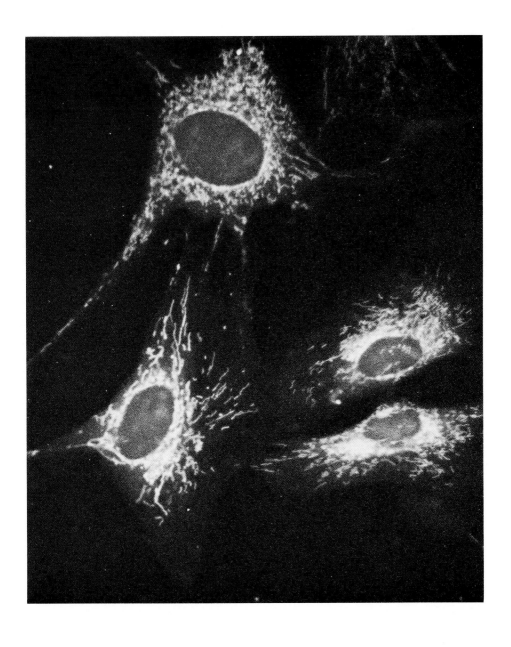

—————— **PLATE 35** ——————

Localization of a Lysosomal Enzyme in Lysosomes. This image shows a rabbit fibroblast in culture labeled using an affinity-purified antibody to rabbit cathepsin D (32). The pattern shown represents lysosomes in the cytoplasm; the enzyme appears to be uniformly distributed in the lysosome. (Mag., ×1400; N, nucleus.)

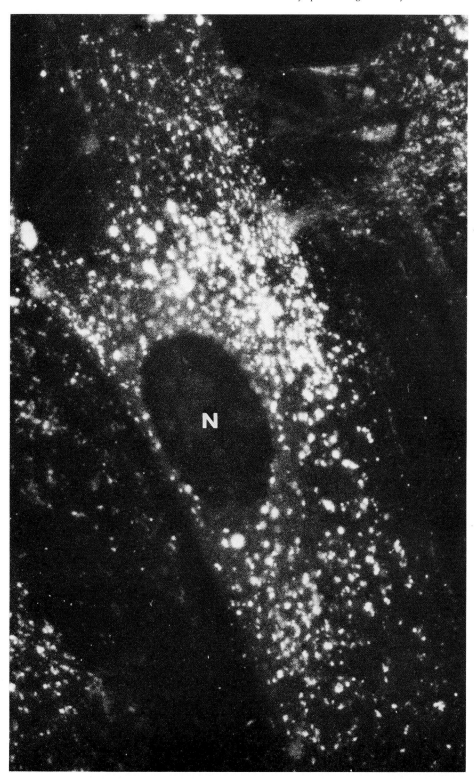

——————— PLATE 36 ———————

Localization of a Lysosomal Membrane Protein. This image shows NIH3T3 cells at high density, labeled using a monoclonal antibody (1D4B) to a lysosomal membrane protein (a gift from Dr. J. T. August) (J. W. Chen, T. L. Murphy, M. C. Willingham, I. Pastan, and J. T. August, manuscript in preparation). The pattern shown demonstrates the intense perinuclear accumulation of lysosomes in these cells. On close examination, the lysosomal pattern can be seen to represent the lysosomal surface and not the lumen, producing images that appear as small "doughnuts" in the cytoplasm. (Mag., ×1300.)

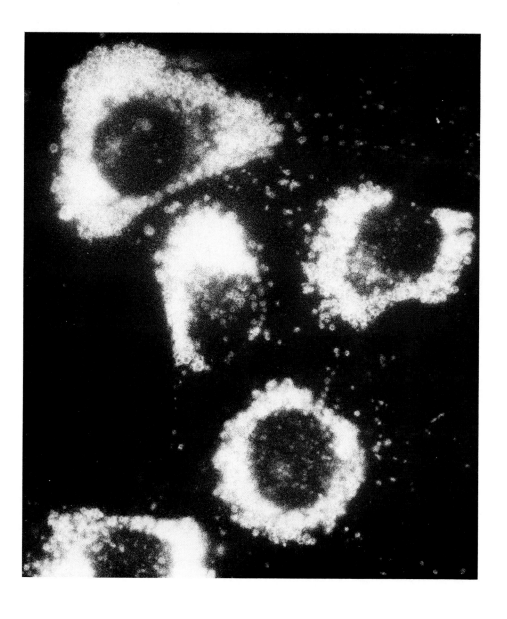

———————— PLATE 37 ————————

Lysosomes Labeled by an Endocytosed Ligand. This image shows Swiss 3T3 cells that have taken up α_2-macroglobulin by endocytosis (30). This ligand is taken into the cell by receptor-mediated endocytosis and is finally concentrated in lysosomes, where it is slowly degraded. These cells have then been labeled after fixation using an antibody to α_2-macroglobulin (48). The pattern shown represents the accumulation of α_2-macroglobulin in the lumen of each lysosome. (Mag., ×1150; N, nucleus.)

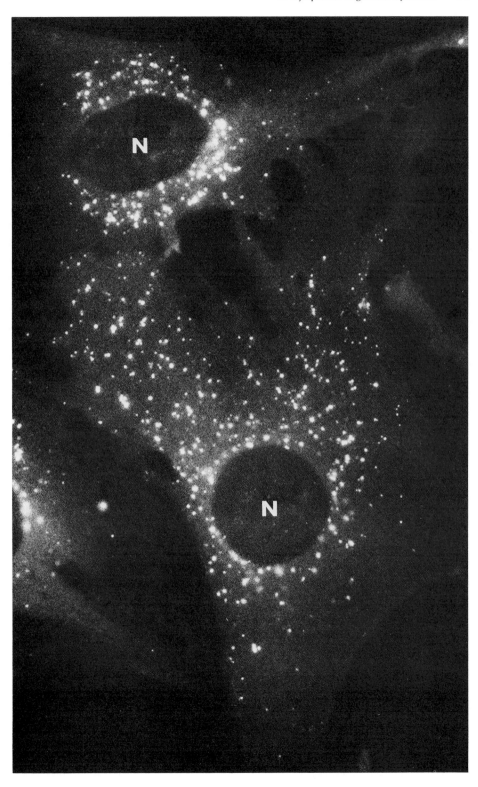

———————— PLATE 38 ————————

Localization in the Endoplasmic Reticulum and Nuclear Envelope. This image shows NIH3T3 cells labeled using a monoclonal antibody that selectively labels the endoplasmic reticulum and nuclear envelope (a gift from Dr. J. T. August) (manuscript in preparation). The nuclear envelope labeling shows a diffuse increase in brightness of the nuclear region that can be seen to be at the nuclear surface by the bright ring at the edge of the nucleus, as well as by focusing up and down through the nuclear image. Note that no sparing of the nucleoli can be detected, indicating that the localization is not within the nuclear matrix. The endoplasmic reticulum appears as a diffusely distributed reticular pattern in the cytoplasm, showing some sparing of other large organelles, such as lysosomes and lipid droplets. Since the endoplasmic reticulum and nuclear envelope are contiguous, this would be the pattern expected for an antigen present in this organelle. (Mag., ×1350.)

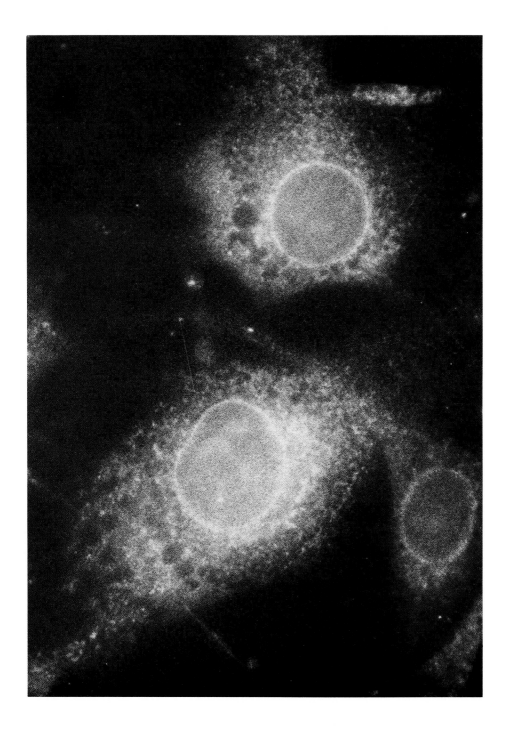

—————— PLATE 39 ——————

Transfer of Protein from Endoplasmic Reticulum to Golgi Stacks. This plate represents a functional kinetic experiment using the O-45 mutant of the vesicular stomatitis virus (VSV) (5, 45). With this mutant, when the infected Swiss 3T3 cells are maintained at 39.8°C (the nonpermissive temperature), the "G" protein of VSV is retained within the endoplasmic reticulum. In (A), the "G" protein has been localized in cells fixed from this nonpermissive temperature using anti-"G" antibodies. When cells are shifted to the permissive temperature (32°C), the "G" protein is released from the endoplasmic reticulum and proceeds into the Golgi stacks, where it can be seen using anti-"G" antibodies 12 minutes after temperature shift (B). This experiment functionally defines the location of the endoplasmic reticulum and nuclear envelope (A) and the Golgi stack system (B). Note that the stacks (arrow) are confined to a small region just adjacent to the nucleus (N). The location of this antigen in these sites has been confirmed by electron microscopic immunocytochemistry. (Mag., ×1600.)

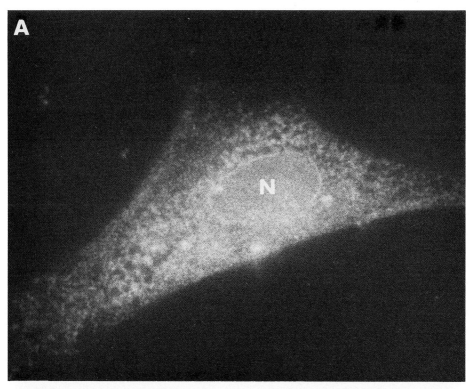

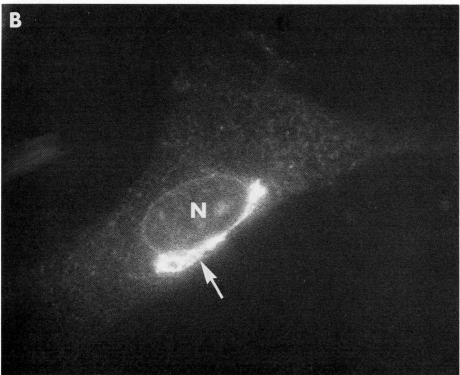

————————— PLATE 40 —————————

Golgi Stack Labeling in Rounded Cells. These images show labeling of Golgi stacks in Harvey NIH3T3 transformed cells using a monoclonal antibody (ABL 70) to a protein antigen that localizes to the stacks of the Golgi by electron microscopic immunocytochemistry (a gift from Dr. J. T. August) (manuscript in preparation). Note that the Golgi stacks are concentrated in a relatively small area near the edge of the nuclei (N). Sometimes this perinuclear region image is displayed over or under the image of the nucleus in these rounded cultured cells. (Mag., ×1250.)

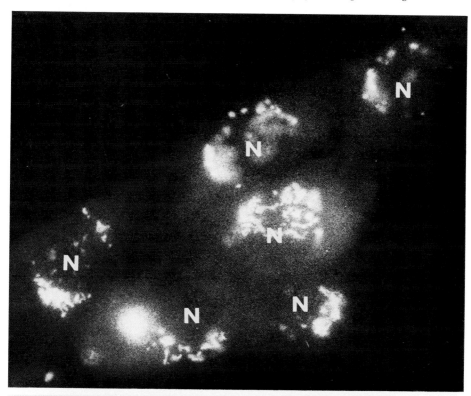

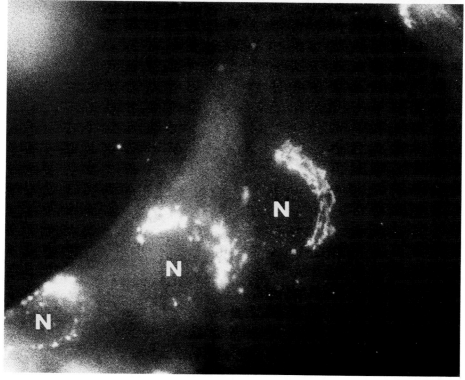

——————— PLATE 41 ———————

Golgi Stack Labeling in Flattened Cells. These images show flattened, untransformed NIH3T3 cells labeled using a monoclonal antibody to a protein present in the Golgi stacks (ABL 70) (a gift from Dr. J. T. August) (manuscript in preparation). The display of the stacked cisternae shows a highly confined distribution just adjacent to the nuclear envelope. On occasion, the stacks are located over or above the image of the nucleus in these cultured cells; some of the extensive arrays of stacks with this distribution are shown in these fields. All of the labeled structures shown in these images are directly adjacent to the nuclei. (Mag., ×1350.)

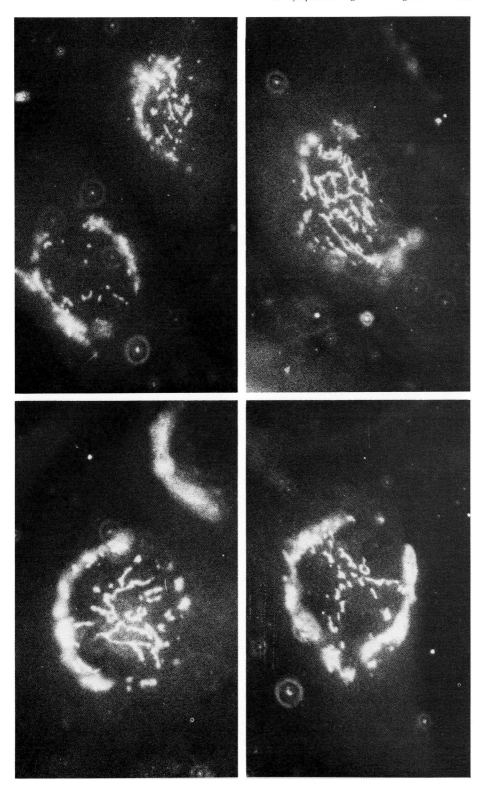

——————— PLATE 42 ———————

Transreticular Golgi Network. This image shows a chick embryo fibroblast labeled using anti-clathrin to demonstrate the coated regions present in the transreticular network of the Golgi (45, 50). Note the perinuclear array that extends further into the cytoplasm, which in one area appears as a separated reticular network (arrow). This array corresponds to the reticular elements of the Golgi system that contains clathrin-coated regions; the plasma membrane-coated pits are the small punctate dots seen over the remainder of the cell image. (Mag., ×1350; N, nucleus.)

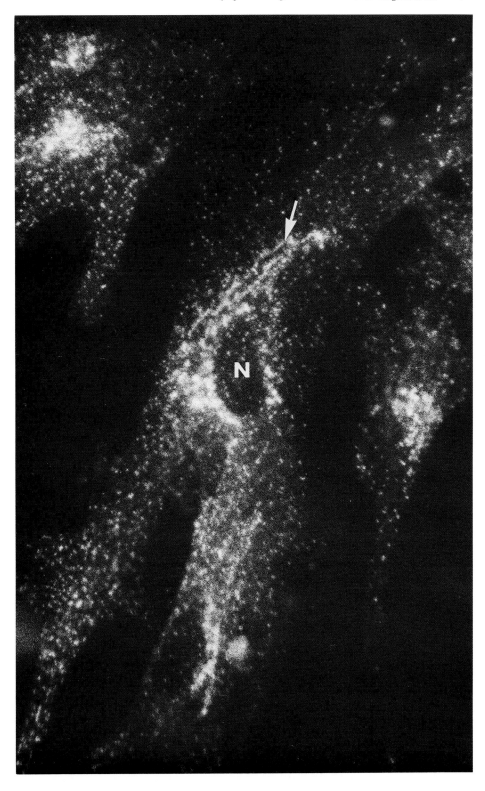

———————— PLATE 43 ————————

Transreticular Golgi Network. These images show human fibroblasts (Detroit 551) labeled using an antibody to the phosphomannosyl receptor (a gift from Dr. G. G. Sahagian) (54), which show elaborate displays of the transreticular networks of the Golgi system. These very flattened cells stretch this network into a more elongated horizontal plane, allowing individual tubular elements of the network to be clearly visible. The presence of this receptor in such elements has been confirmed by electron microscopic immunocytochemistry. Some of this pattern may reflect a more complex distribution of the adjacent Golgi stacked cisternae that occurs in such flattened primary cell lines. (Mag., ×1150; N, nucleus.)

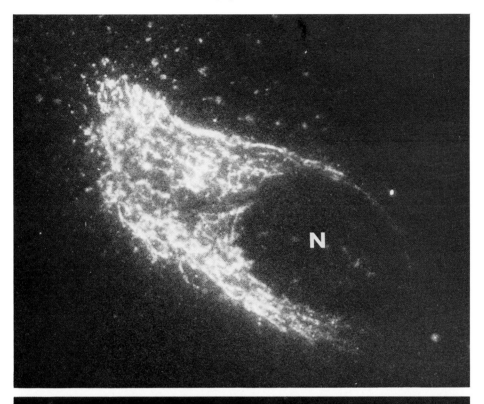

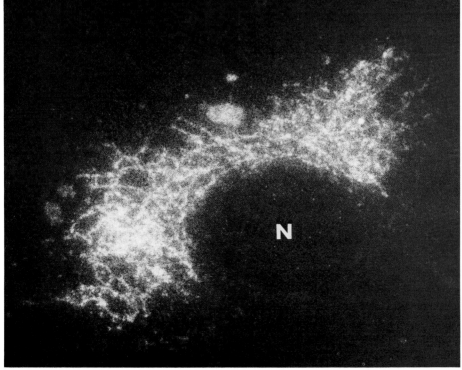

———————— PLATE 44 ————————

Transreticular Network in Elongated Cells. These images show elongated human fibroblasts labeled using an antibody to the phosphomannosyl receptor (a gift from Dr. G. G. Sahagian) (54). Note that the transreticular network can extend far away from the perinuclear region, stretching out cell processes for great distances. (Mag., ×1150; N, nucleus.)

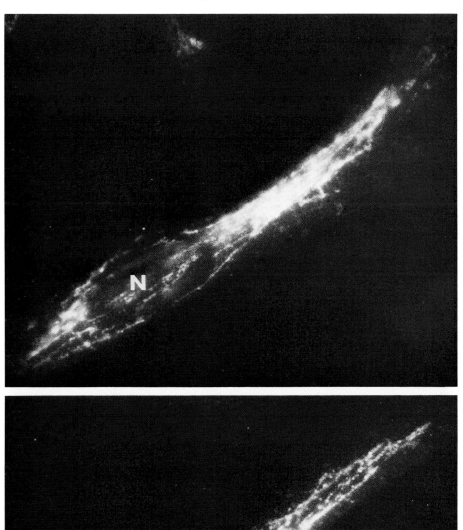

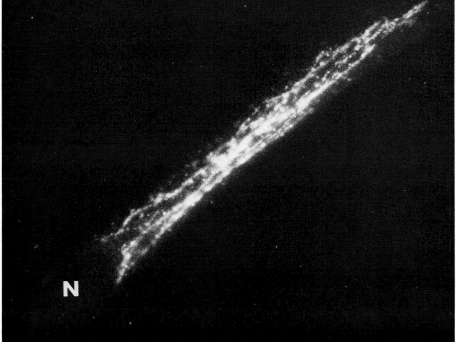

───────── PLATE 45 ─────────

Transreticular Golgi Labeled Using Anti-Transferrin Receptor. This image shows a human fibroblast (Detroit 551) labeled using a monoclonal antibody (HB 21) to the human transferrin
receptor. In this well-spread fibroblast one can see the labeling by this antibody of the
transreticular network spreading out from the perinuclear region. By electron microscopy, this
antibody does not label the Golgi stacks, but only labels the transreticular elements, endocytic
vesicles (receptosomes), and plasma membrane-coated pits. These surface-coated pits can be
seen as the finely punctate structures over the cell surface in this image. (Mag., ×1250; N,
nucleus.)

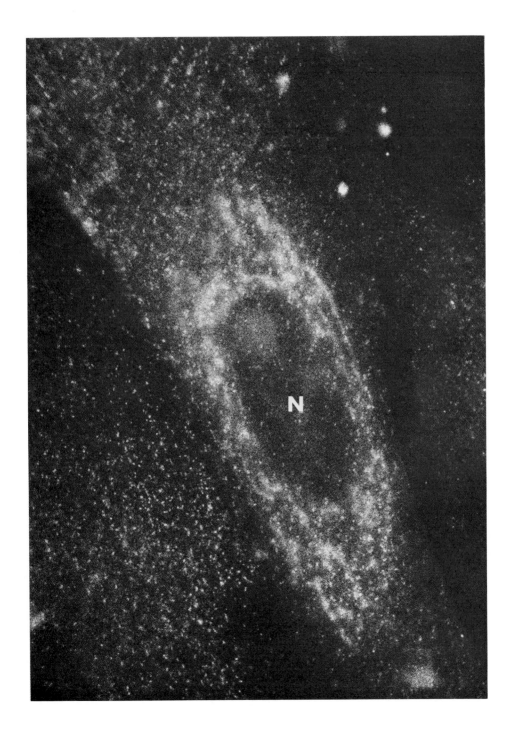

———————— **PLATE 46** ————————

Labeling of the Golgi Region. This image shows a Swiss 3T3 cell labeled using a monoclonal antibody (2C6) that reacts with the α_2-macroglobulin receptor (a gift from Dr. J. T. August) (manuscript in preparation). This receptor appears on the cell surface but mainly is present in endocytic vesicles (receptosomes) and trans-Golgi elements. This image shows the accumulation of label in the perinuclear Golgi region, in addition to the punctate endocytic vesicle labeling in the cell periphery. (Mag., ×1250; N, nucleus.)

——————— PLATE 47 ———————

Endocytic Vesicle Labeling. This image shows a Swiss 3T3 cell labeled with α_2-macroglobulin conjugated to rhodamine (30). This cell was incubated for 10 minutes at 37°C, at which time this ligand had been delivered from surface-coated pits into endocytic vesicles (receptosomes). The cell was then fixed in formaldehyde; no antibodies were used. This pattern shows the display of endocytic vesicles in this type of functional experiment. Note that many of the vesicles are randomly distributed in the cytoplasm, but some are beginning to accumulate in the perinuclear Golgi region. (Mag., ×1300; N, nucleus.)

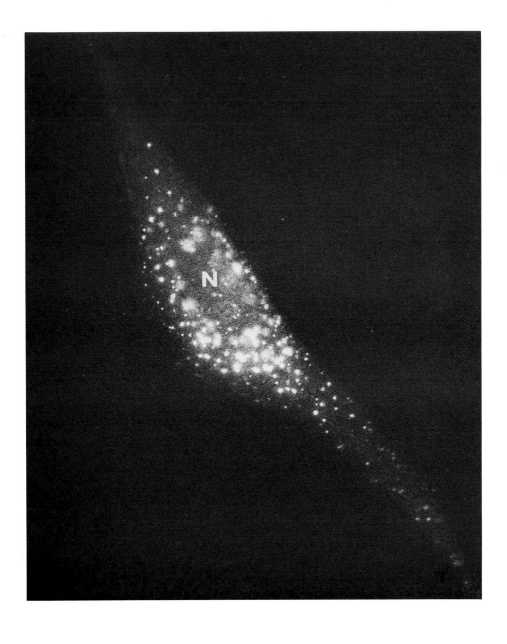

———————— PLATE 48 ————————

Endocytic Vesicle Labeling in a Flattened Cell. This image shows the display of the α_2-macroglobulin receptor in a flattened Swiss 3T3 mouse fibroblast using a monoclonal antibody (2C6) (a gift from Dr. J. T. August) (manuscript in preparation). The many punctate structures randomly distributed throughout the cytoplasm represent endocytic vesicles (receptosomes) that contain receptor. Some receptor is also present in coated pits at the cell surface. α_2-Macroglobulin is a component of the culture medium and is being constantly internalized along with its receptor. A concentration of receptor-containing elements can also be seen in the Golgi region. (Mag., ×1250.)

—————— PLATE 49 ——————

Mixed Organelle Pattern. This image shows the distribution of the transferrin receptor in KB cells in interphase (A) and during mitosis (B) using a monoclonal antibody (HB 21). (C) is a phase-contrast image of the cells in (B). In (A), the receptor is distributed at the cell surface, in endocytic vesicles, and in the trans-Golgi region, generating the mixed pattern shown. In mitosis (C), the receptor-containing membranes aggregate around the pericentriolar region on either side of the mitotic spindle (arrows) in both metaphase (M) and telophase (T) cells. (Mag., ×1700; N, nucleus.)

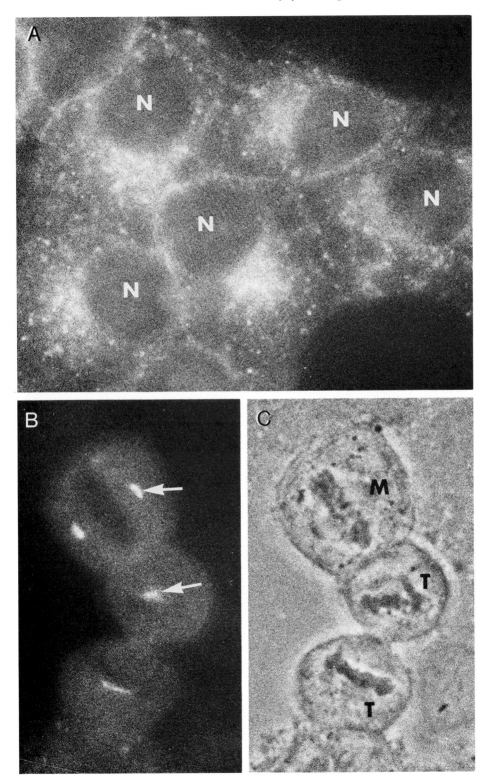

—————————— PLATE 50 ——————————

Diffuse Microfilament Meshwork. This image shows the diffuse microfilament meshwork just beneath the plasma membrane in a flattened Swiss 3T3 cell using antibodies to fodrin (nonerythrocyte spectrin) (a gift from Dr. C. B. Klee) (25). Note that there are dark lines that run through the image at the periphery of the cell; these represent microfilament bundles which do not label with anti-fodrin. Note also that the bright meshwork lies over and under the image of the nucleus (N). (Mag., ×1350.)

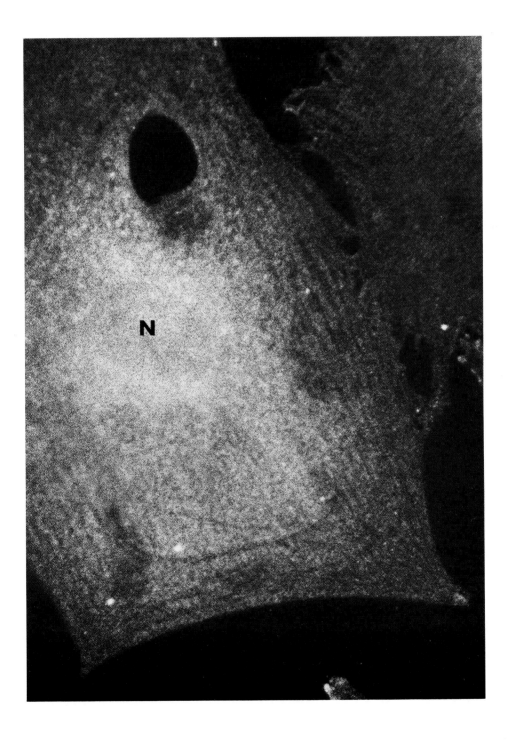

——————— PLATE 51 ———————

Diffuse Actin Meshwork. This image shows a chick embryo cell labeled using anti-actin anti-bodies (24, 58). Note the diffuse actin meshwork that overlies the entire cell image. The lack of nuclear image sparing shows that the localization is not deep in the cytoplasm but is just beneath the plasma membrane on the top and bottom of the image of the nucleus (N). The meshwork is contiguous with the microfilament bundles at the cell periphery which label with anti-actin. (Mag., ×1200.)

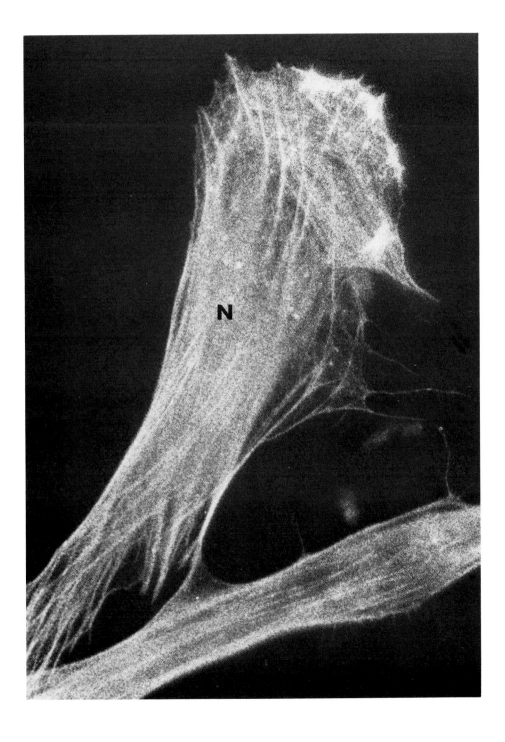

——————— PLATE 52 ———————

Microfilament Meshwork: Anti-Alpha Actinin. This image shows the localization of alpha-actinin using a monoclonal antibody in chick embryo fibroblasts (23, 43). Note the diffusely punctate, interrupted labeling of the meshwork and microfilament bundles characteristic of alpha-actinin. The labeling of bundles, ruffles, and the interrupted nature of the localization distinguish this from other microfilament-associated proteins such as actin (which would not show periodicity), myosin (which would not label ruffles or lamellae), or tropomyosin (which would not label ruffles or lamellae). (Mag., ×1150.)

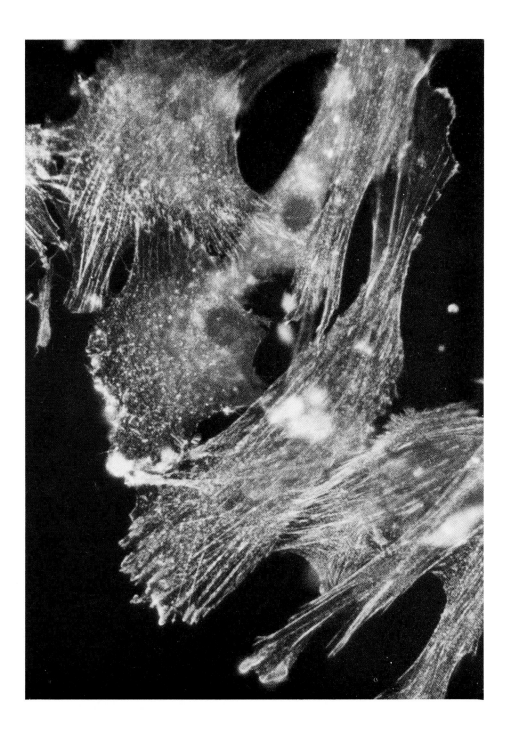

———————— PLATE 53 ————————

Focal Aggregations in the Microfilament Meshwork: Myosin. This image shows a Swiss 3T3 mouse fibroblast labeled with anti-myosin (40, 59). Note that the association of myosin with the microfilament bundles shows a marked periodicity, and the localization does not extend far into the lamellae. At the edge of leading lamellae, the myosin forms focal aggregations in the meshwork with large spacing between the periodic deposits. Note also that little or no myosin is evident in the periphery of the lamellae, in marked contrast to the large amounts of actin and alpha-actinin in these regions. (Mag., ×1250; N, nucleus.)

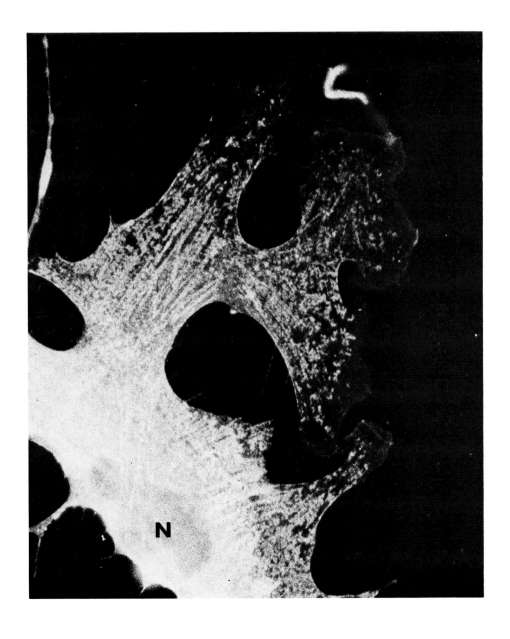

——————— PLATE 54 ———————

Actin Localization in a Migrating Fibroblast. This image shows a Swiss 3T3 cell labeled using anti-actin antibodies (24, 58). This cell has a leading lamella indicative of cell migration, as well as some small microfilament bundles stretching longitudinally down the extending cell process. These bundles label uniformly with anti-actin. The periphery of the lamella shows a concentration of actin (arrowhead on frilly edge), and smaller surface aggregations can be seen just behind this leading edge (arrow) which probably represent the remnants of upper surface ruffle activity or microvilli. (Mag., ×1300; N, nucleus.)

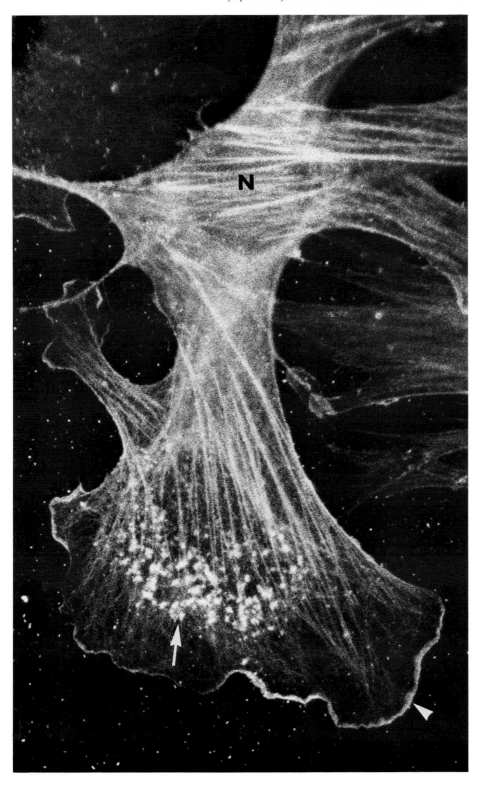

———————— PLATE 55 ————————

Microfilament Bundles: Alpha-Actinin. This image shows the display of microfilament bundles in a spread chick embryo fibroblast using antibodies against alpha-actinin (23, 43). The bundles extend from the peripheral edges of the extended cell and insert into the lower cell surface at focal adhesions. The alpha-actinin pattern is characterized by a periodic arrangement along the bundles and an extension of the label into the lamellar edges (arrow). (Mag., ×1350; N, nucleus.)

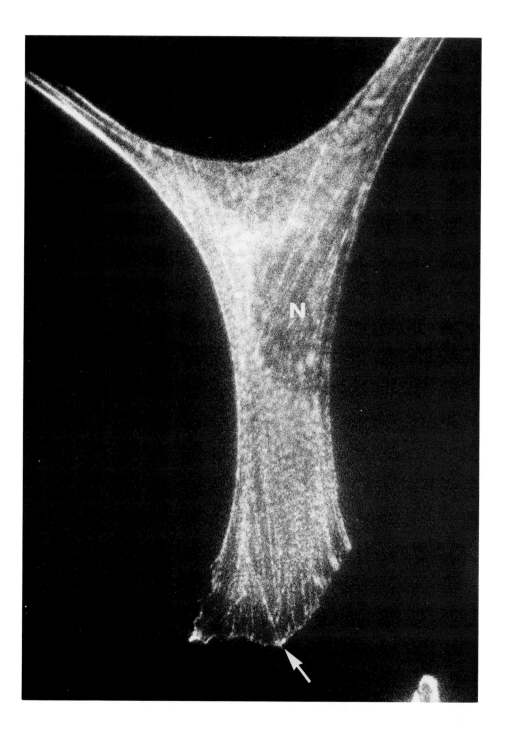

PLATE 56

Microfilament Bundles: Anti-Actin. This image shows relatively small Swiss 3T3 cells spread on a substrate and labeled using anti-actin (24, 58). The microfilament bundles are prominently labeled. They usually stretch from the edges of the cell periphery on the lower substratum surface, although occasionally they can be found under the upper cell surface. The labeling with anti-actin shown here is completely uniform along the length of the bundles. The center of these cells appears to show bundles that are out of focus, yet they are in focus at the cell periphery. Assuming that the substratum is flat, this observation suggests that these bundles are following the upper cell surface in this field; this is because the bulge of the cell in the nuclear region generally extends away from the plane of the substrate. This interpretation could be confirmed by focusing up and down on the cell. (Mag., ×1150.)

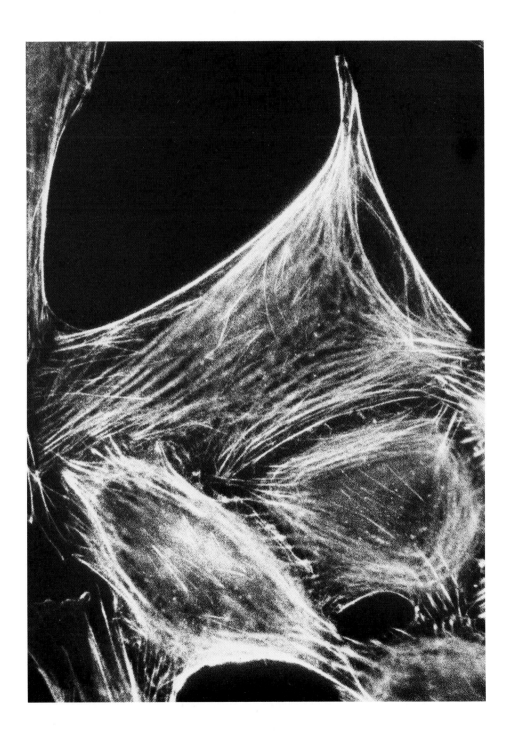

——————   PLATE 57   ——————

Microfilament Bundles in an Elongated Cell: Anti-Actin. This image shows an extensively elongated Swiss 3T3 cell labeled using anti-actin (24, 58). Note the prominent microfilament bundles that lie parallel to the long axis of the cell. The labeling with anti-actin is homogeneous and uniform along the bundles. (Mag., ×1150.)

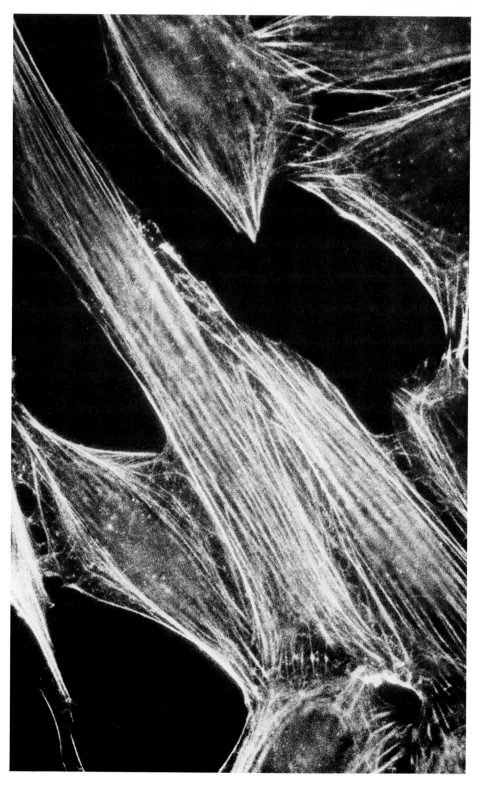

——————— PLATE 58 ———————

Microfilament Bundles in a Flattened Cell: Anti-Actin. This image shows a portion of a highly flattened Swiss 3T3 cell labeled using anti-actin (24, 58). The variation in distribution and appearance of microfilament bundles is clearly shown. The entire cell image shows a diffuse, weak labeling which represents the microfilamentous diffuse meshwork. Bundles in this cell lie on the lower cell surface next to the substratum and are oriented in the directions of cell tension exerted on the substratum (hence the term *stress fibers* for microfilament bundles). It is important to note that all of these bundles lie closely apposed to a cell surface, that is, the plasma membrane, in either the center of the cell image or at the cell periphery. (Mag., ×1150.)

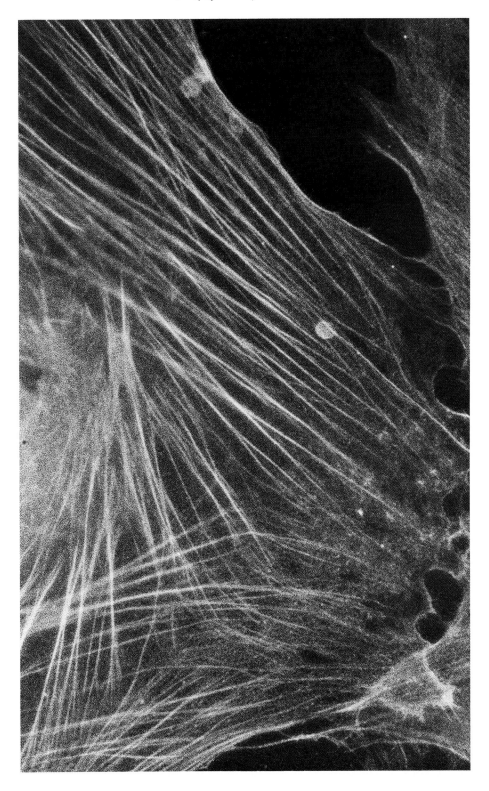

—————— PLATE 59 ——————

Microfilament Bundles: Anti-Myosin. This image shows a portion of a flattened Swiss 3T3 cell labeled using anti-myosin (nonmuscle, fibroblast myosin) (40, 59). This shows multiple micro-filament bundles labeled in the characteristic interrupted periodicity seen for myosin associated with stress fibers in flattened cells. In cells that are less flattened, this periodicity may not be so evident. In one portion of this cell the distribution along some bundles almost appears uniform (arrow). (Mag., ×1150.)

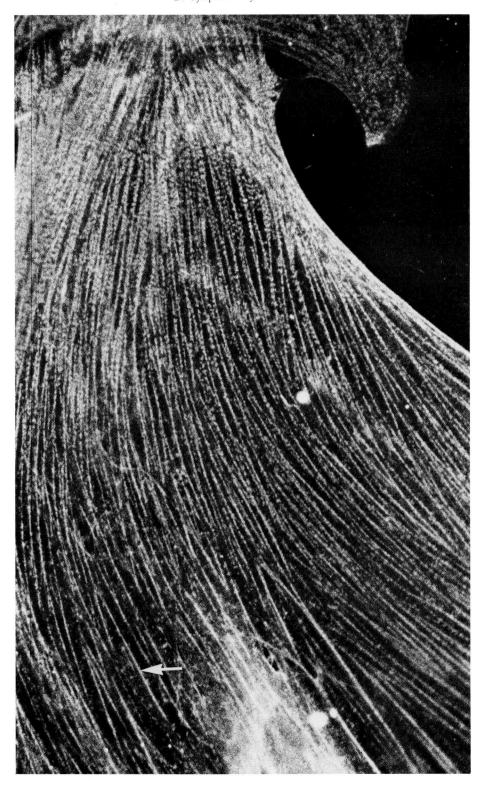

PLATE 60

Microfilament Bundles: Anti-Tropomyosin. This image shows a small chick embryo fibroblast labeled using anti-tropomyosin antibodies (22). The small microfilament bundles are labeled in a uniform pattern with this antibody with only a slight hint of periodicity, but labeling does not extend into the lamellae at the cell periphery, distinguishing this from anti-actin. (Mag., ×1250.)

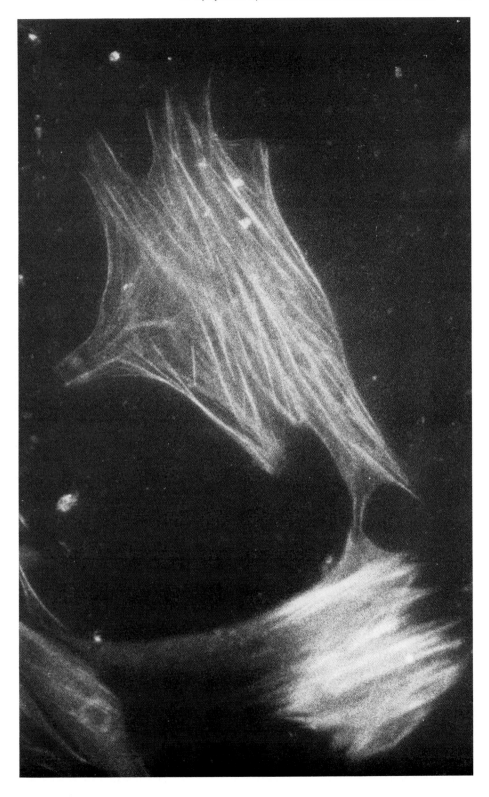

———————— PLATE 61 ————————

Microfilament Bundles in an Elongated Cell: Anti-Tropomyosin. This image shows a very elongated chick embryo fibroblast labeled using anti-tropomyosin antibody (22). The large microfilament bundles present in this cell are heavily labeled with anti-tropomyosin. Without the decreased labeling at the cell periphery and the very occasional periodicity, this pattern would be indistinguishable from that of anti-actin. (Mag., ×1250.)

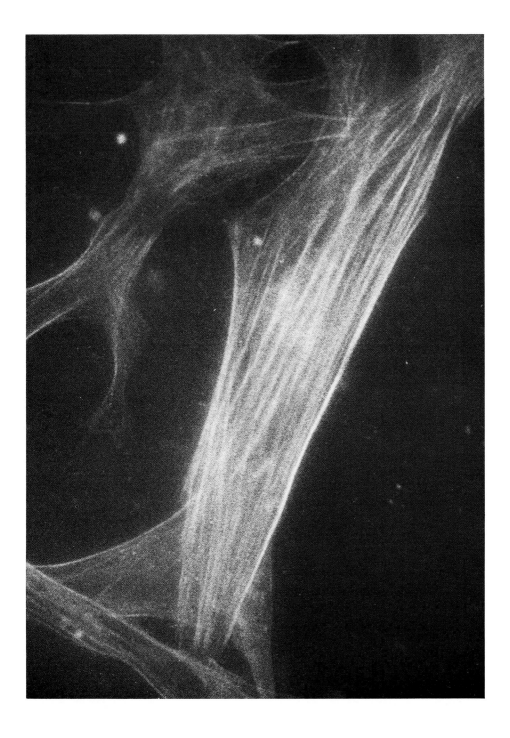

——————— PLATE 62 ———————

Microtubules: Glutaraldehyde–Borohydride Fixation and Anti-Tubulin. This image shows a Swiss 3T3 cell that has been fixed using glutaraldehyde followed by a sodium borohydride treatment prior to labeling with anti-tubulin (42, 48). This fixation protocol preserves microtubules exceptionally well, and one can see individual microtubules, especially under the nuclear image. A site of confluence of microtubules, the microtubule organizing center (MTOC), lies adjacent to the nucleus (arrow). Individual microtubules stretch out from this site into the cell periphery. (Mag., ×1300.)

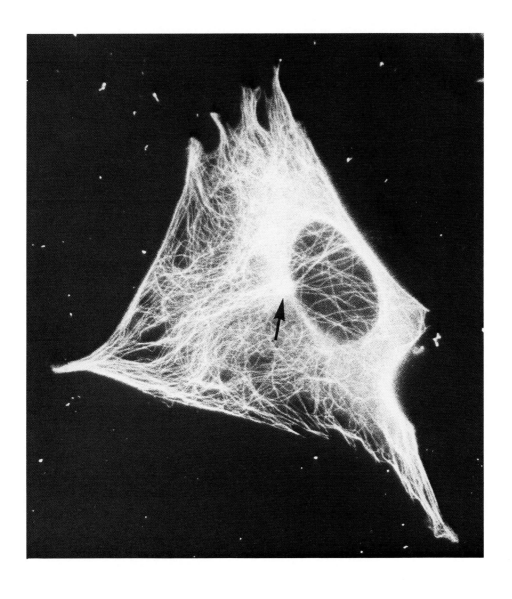

————— PLATE 63 —————

Microtubules: Peripheral Terminations Using Anti-Tubulin. This image shows a Swiss 3T3 cell fixed and processed using the glutaraldehyde–borohydride method and anti-tubulin (42, 48). In this cell, the terminations of individual microtubules in the cell periphery can be clearly seen (small arrows). (Mag., ×1250.)

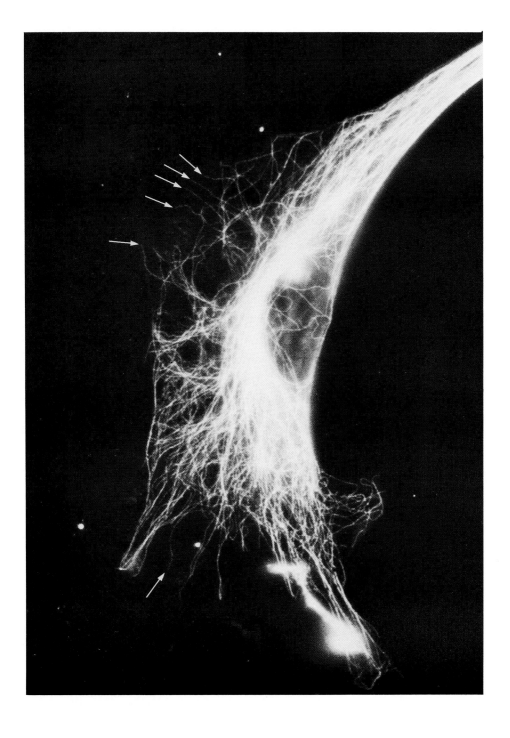

────── PLATE 64 ──────

Microtubules: Flat Cell Using Anti-Tubulin. This image shows a flat Swiss 3T3 cell fixed using glutaraldehyde followed by borohydride and labeled with anti-tubulin (42, 48). In such a flat cell the display of most of the microtubules in the cell becomes clearer. While these tubules extend out into the thin lamellae at the cell periphery, they are usually separated from the plasma membrane by the actin microfilamentous meshwork. Note that while the tubules extend out toward the periphery, this has no relationship to their spacing from the plasma membrane, which lies on the top and the bottom of the cell. (Mag., ×1250.)

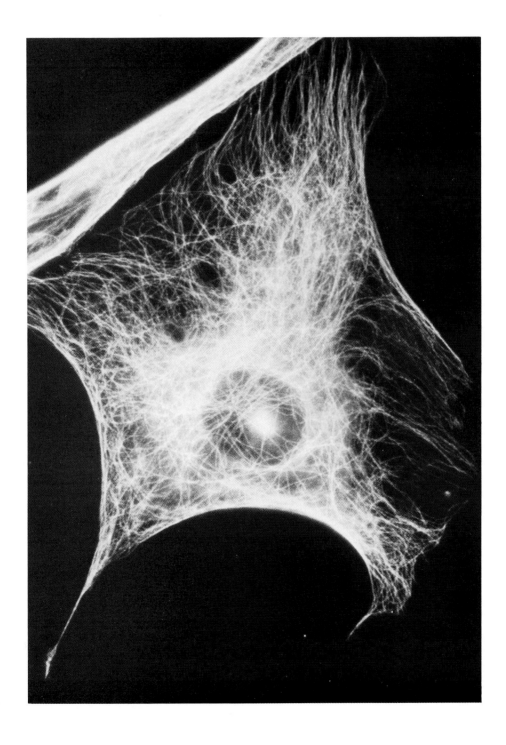

———————— PLATE 65 ————————

Microtubules: Anti-Tubulin in a Very Flat Cell. This image shows a very flat Swiss 3T3 cell labeled using anti-tubulin (42, 48). In such a flat cell, almost all of the microtubules become visible. The large size of this image is caused by cell flatness, since the actual overall volume of the cell is similar to that of a more rounded cell. Thus, the cell with this large area on the substratum is extremely thin, in some cases as thin as 1–2 μm. (Mag., ×1250; N, nucleus.)

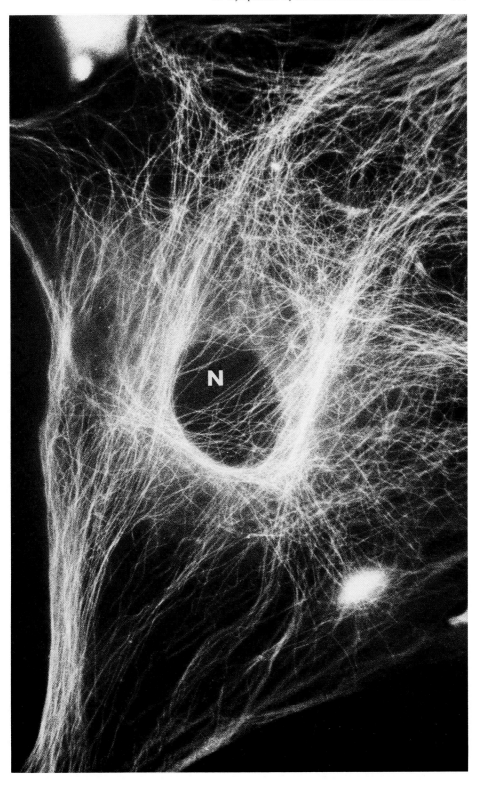

————— PLATE 66 —————

Microtubule Organizing Center: Anti-Tubulin. This image shows Swiss 3T3 cells fixed using formaldehyde that are labeled with anti-tubulin. The image is out of focus to emphasize the high concentration of label in the perinuclear region near the centrioles and the MTOC (arrows) (6). These bright perinuclear regions are the point of confluence of microtubules in the cell periphery. (Mag., ×1350.)

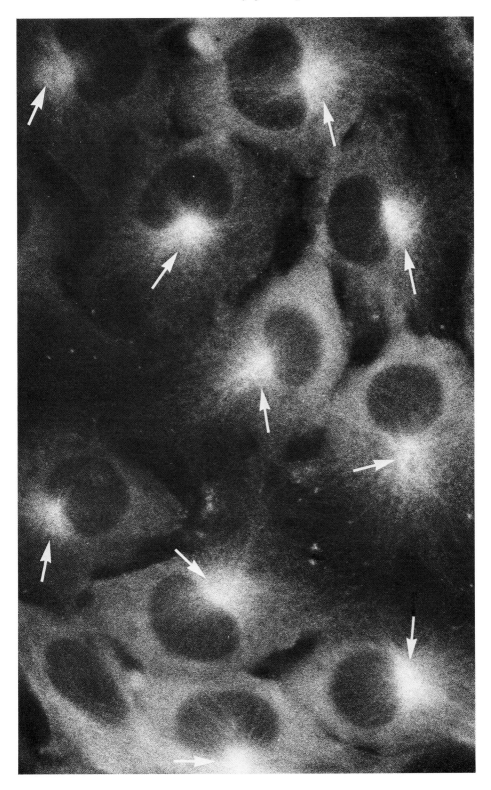

───────── **PLATE 67** ─────────

Primary Cilia of Flattened Fibroblasts Using Anti-Tubulin. This image shows two primary cilia (arrows) visualized in flat Swiss 3T3 cells using anti-tubulin antibody. These cilia originate from the centrioles in the perinuclear region of many cells and are observed during certain parts of the cell cycle. Their visualization is often made easier when they lie projecting over the nuclear image, which is dark and provides a clearer background. They are difficult to see in many cells. (Mag., ×1700.)

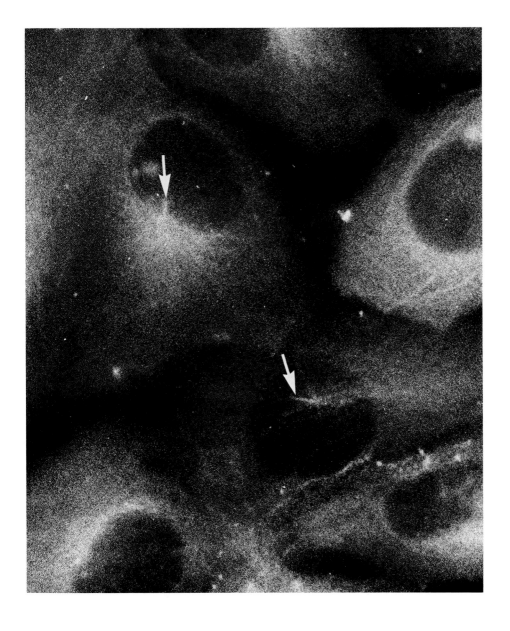

—————— PLATE 68 ——————

Intermediate Filaments: Vimentin. This image shows a Swiss 3T3 fibroblast labeled using antibodies to vimentin, the single intermediate filament protein present in these cells (7, 12). The pattern is characteristically a wavy collection of filaments that extend toward the cell periphery. These can most easily be distinguished from tubulin by examining mitotic cells, in which anti-tubulin will specifically label the mitotic spindle, while this antibody will not. The overall pattern in interphase cells is reminiscent of microtubules, since these intermediate filaments often accompany microtubules through their paths to the cell periphery. (Mag., ×1350.)

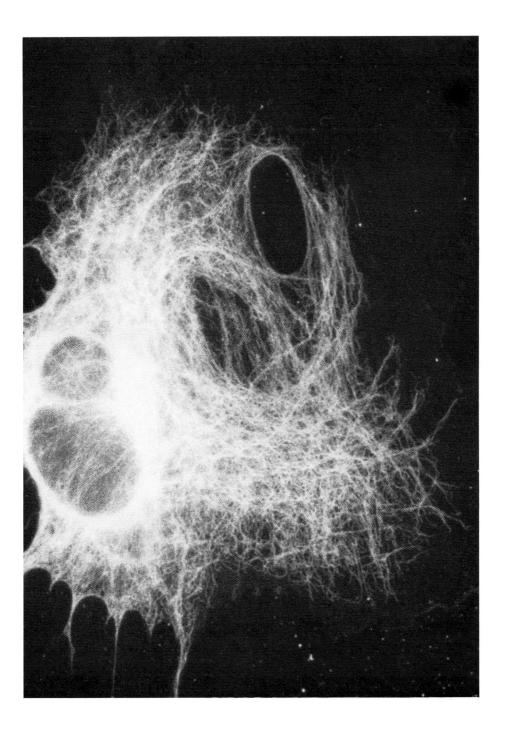

———————— PLATE 69 ————————

Intermediate Filaments: Cytokeratin. This image shows a flat human primary mesothelial cell (a gift from Dr. S. P. Banks-Schlegel) labeled using an antibody to keratin (12, 29). These epithelial cells contain large numbers of keratin filaments in bundles in the cytoplasm, which characteristically wrap around nuclei as shown in this binucleate cell. (Mag., ×1000; N, nucleus.)

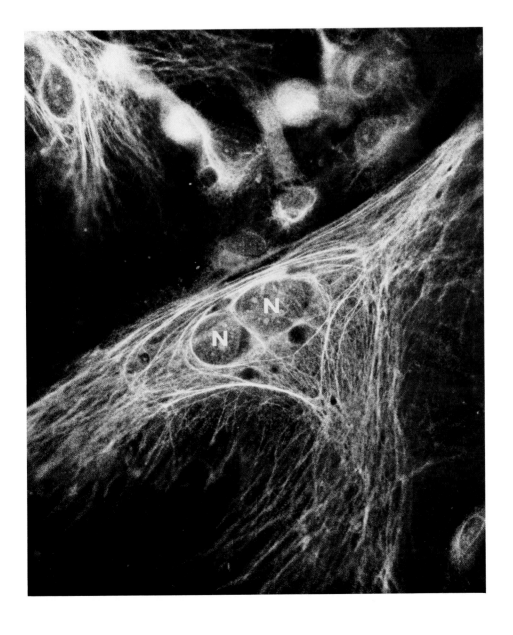

———————— PLATE 70 ————————

Paracrystalline Tubulin Induced by Vinblastine Treatment. This image shows Swiss 3T3 cells treated for 1 hour with vinblastine (100 μM) and labeled using anti-tubulin antibodies (38). Note the discrete cytoplasmic inclusions that represent tubulin paracrystalline arrays induced by this drug. (Mag., ×1300; N, nucleus.)

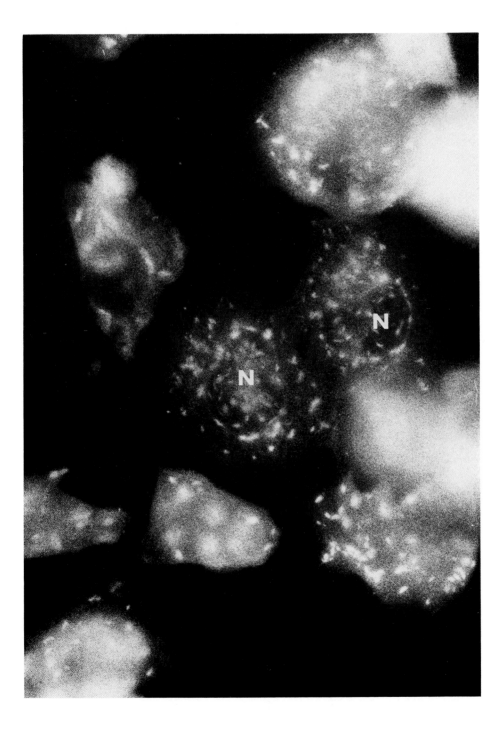

——————— PLATE 71 ———————

Crystalloid Cytoplasmic Inclusions: Nematin. This image shows a group of NRK-cultured rat cells that are labeled using a monoclonal antibody against a unique crystalloid protein termed *nematin* (manuscript in preparation) (36). These inclusions often occur as threads or toroids in the cytoplasm; often only one such structure is present in each cell. By electron microscopy, these structures appear similar to short aggregates of intermediate filaments, but they do not react with antibodies to other cytokeratins or to tubulin. These structures occur at low frequency in many cultured cell lines. (Mag., ×1350; N, nucleus.)

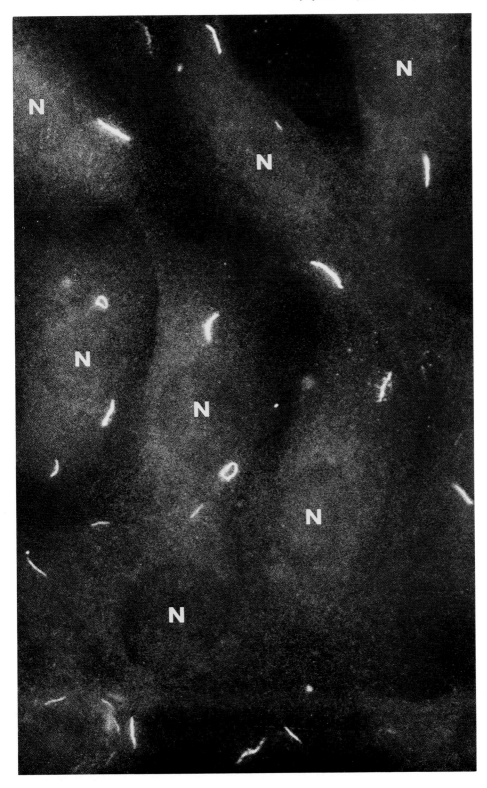

——————— PLATE 72 ———————

T Antigen in the Nuclei of SV40-Transformed Cells. This image shows a group of SV40-transformed BALB 3T3 cells that have been labeled using anti-T antigen antibodies (a gift from Dr. G. W. H. Jay) (33). This antigen is present in the nuclear matrix in SV40-transformed cells and is excluded from the nucleoli, which appear as small dark regions within the nuclear image. (Mag., ×700.)

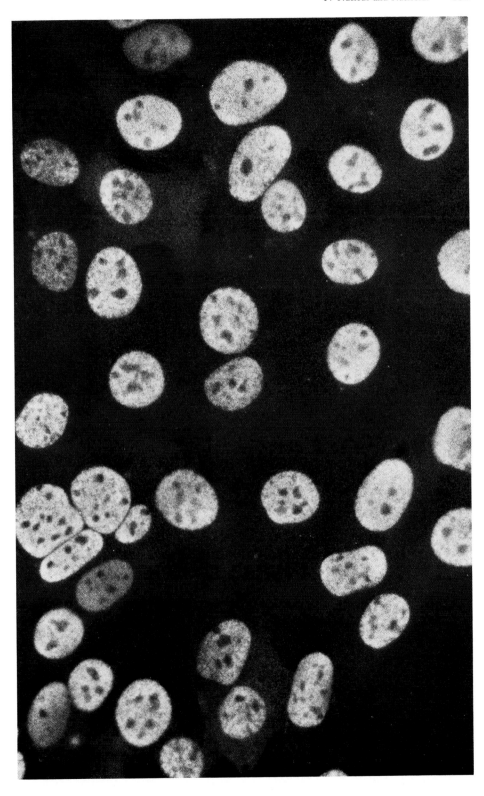

————— PLATE 73 —————

Nuclear Concentration of an Antigen in Transformed Cells. This image shows SV40-transformed BALB 3T3 fibroblasts labeled using an antibody to p53, a protein induced in such transformed cells (a gift from Dr. G. W. H. Jay) (10). This protein shows a nuclear matrix pattern with nucleolar sparing essentially identical to SV40-T antigen. Note the cells in mitosis (arrows) in which the protein is distributed throughout the cytoplasm. (Mag., ×700.)

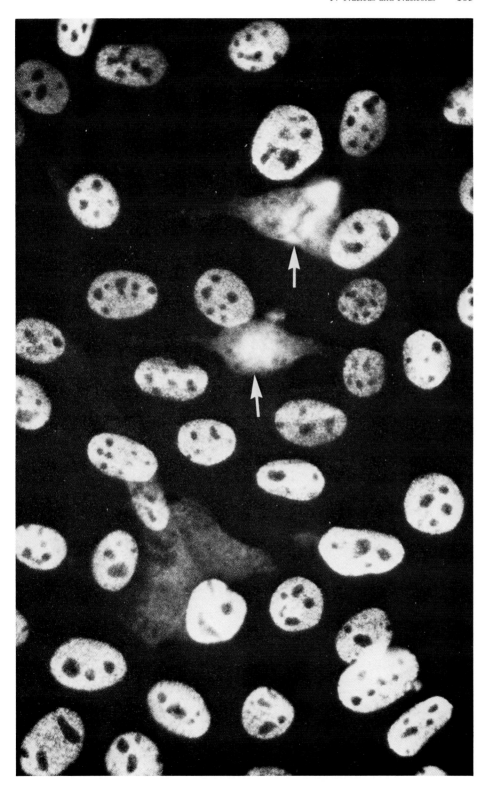

——————— PLATE 74 ———————

Nuclear Antigen in Transformed Cells. These images show SV40-transformed BALB 3T3 cells labeled using a monoclonal antibody to an uncharacterized antigen. This antigen demonstrates a granular nuclear matrix distribution with nucleolar sparing and shows a diffuse cytoplasmic distribution during mitosis (arrow). (Mag., ×1300.)

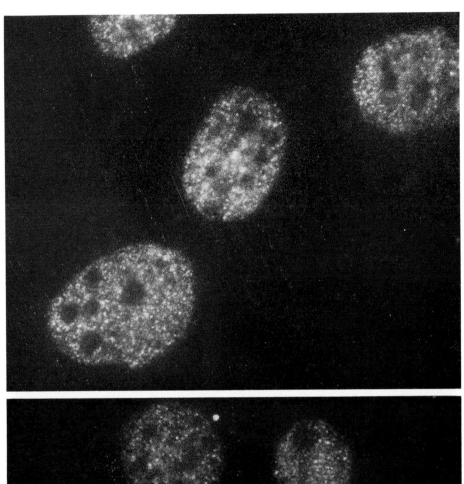

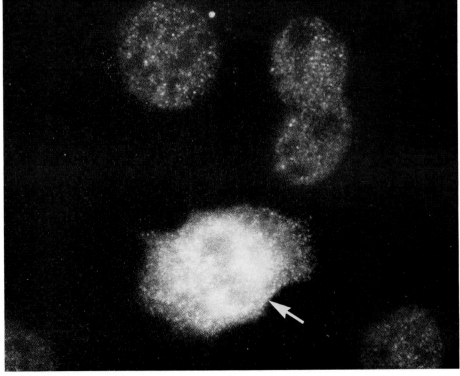

———————— PLATE 75 ————————

Tubulin Localization in Mitosis. These images show the pattern of tubulin distribution (14) in prophase (A) and metaphase (B–D) in mitotic Swiss 3T3 cells. The mitotic spindle labels brightly with anti-tubulin and the dark region in the center of the spindle during metaphase represents the plate of chromosomes aligned at the spindle center that do not label with anti-tubulin. (Mag., ×1250.)

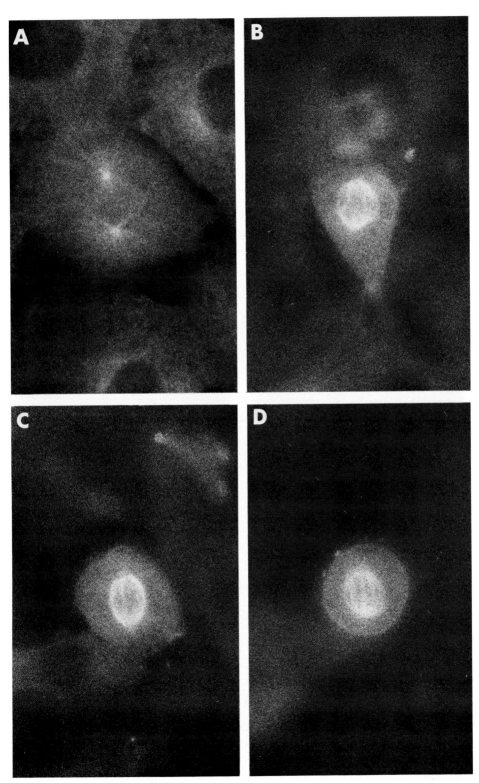

——————— PLATE 76 ———————

Tubulin Localization in Mitosis. These images show the pattern of tubulin localization (14) in anaphase (A,B), early telophase (C), and late telophase (D). The pericentriolar concentration of tubulin in anaphase diminishes, leaving the major concentration of tubulin in late telophase, predominantly at the intercellular bridge (arrows). (Mag., ×1250.)

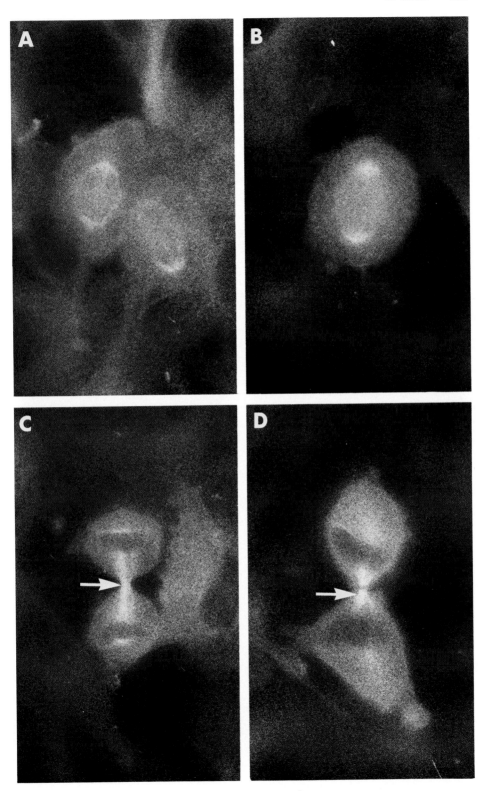

———————— PLATE 77 ————————

Tubulin Localization in the Intercellular Bridge. These images show cells very late in telophase (A) or postmitotic cells (B) labeled using anti-tubulin (14). The intercellular bridge (arrows) shows variable interruptions in tubulin labeling. It always shows a dark region with no labeling that corresponds to the midbody, a site that is so dense that antibodies cannot penetrate. The cell in (A) also shows evidence of surface blebbing activity, a process that usually occurs during cytokinesis due to the constriction of the cleavage furrow. (Mags. A, ×1350; B, ×1250.)

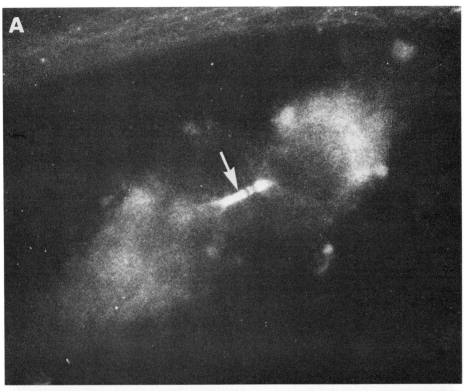

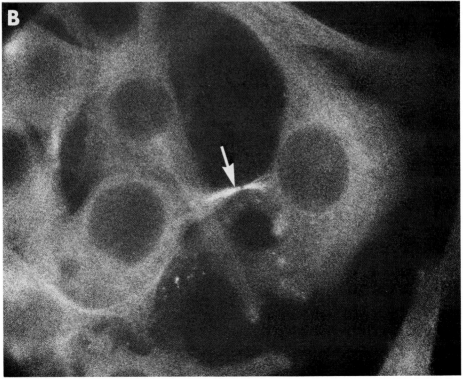

─────── PLATE 78 ───────

Calmodulin Localization in Mitosis. These images show Swiss 3T3 cells labeled using anti-calmodulin antibodies during metaphase (A), anaphase (B), telophase (C), and postmitosis (D). Note the pericentriolar distribution of calmodulin during mitosis, which shifts to a wide intercellular bridge labeling after telophase (1, 46, 57). The width of the empty region in the intercellular bridge seen with anti-calmodulin is much wider than the midbody exclusion region seen using anti-tubulin. Note also that in anaphase and telophase, most of the label for anti-calmodulin is in the pericentriolar region and not in the spindle interzone. (Mag., ×1350.)

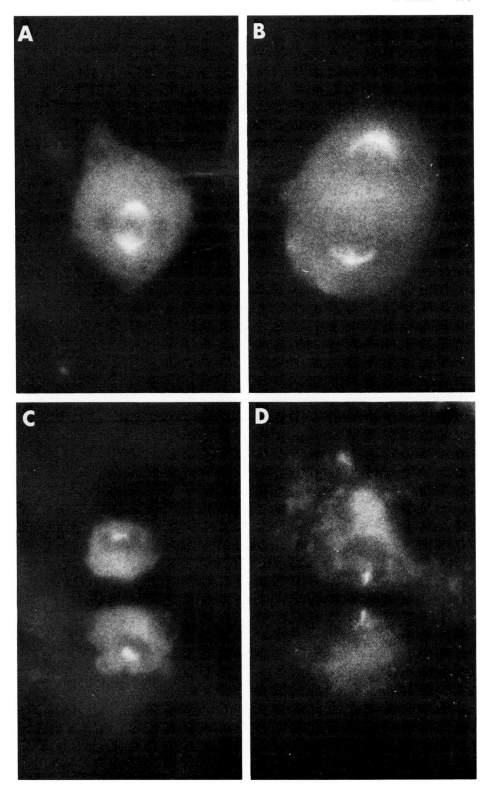

———————— PLATE 79 ————————

Normal Globulin Control. This image shows the level of nonspecific background labeling that should be expected with a properly performed indirect fluorescence labeling procedure. This control involved the substitution of an antibody that did not recognize any antigens in the cell in place of the specific antibody to be used. This indicates that the indirect fluorescently conjugated anti-globulin step shows very little detectable nonspecific binding to the fixed and permeabilized cell. (Mag., ×1250; N, nucleus.)

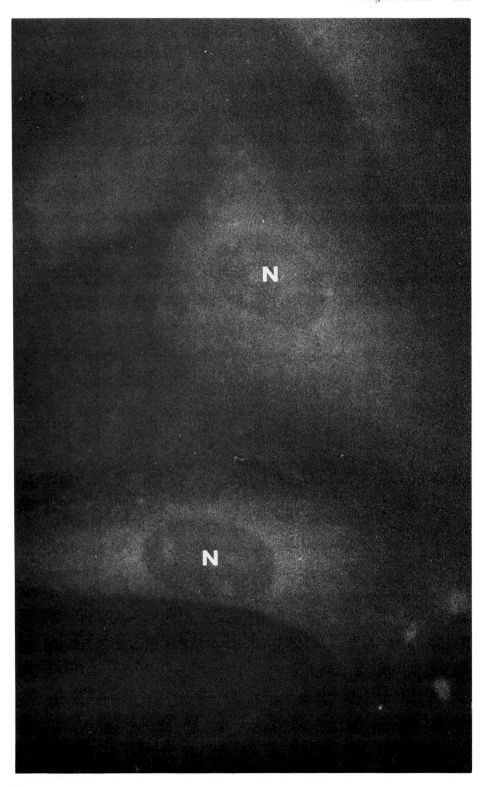

ANTIBODY SOURCES*

Antibody to:	Plate Number
EGF-R1: see ref. 4	5, 6, 8
MC 101: see refs, 13,37	7, 8
YA6-172: see ref. 49	9–13
Clathrin: see ref. 50	14–17, 42
Actin: see ref. 58	18–20, 25, 51, 54, 56–58
Alpha-actinin: Amersham	21, 52, 55
Calmodulin: see ref. 57	22, 23, 28, 78
Tubulin: see ref. 60	23, 62–67, 70, 75–77
Vinculin: Amersham	24
Fibronectin: a gift from Dr. K.Hedman, see ref. 18	26, 27
F1/C5: a gift from Dr. J. T. August	29
CAT: see ref. 17	30
p27: a gift from Dr. N. Richert, see ref. 35	31
F_1-ATPase: a gift from Dr. E. Racker, see ref. 11	32, 33
Cathepsin D: Cappel Labs., M. G. Gallo, unpublished	35
1D4B: a gift from Dr. J. T. August	36
α_2-M: Miles Labs.	37
H69: a gift from Dr. J. T. August	38
"G" protein: see ref. 45	39
ABL 70: a gift from Dr. J. T. August	40, 41
PMR: a gift from Dr. G. G. Sahagian	43, 44
TFR (HB21): American type culture collection	45, 49
2C6: a gift from Dr. J. T. August	46, 48
Fodrin: a gift from Dr. C. B. Klee	50
Myosin: see ref. 59	53, 59
Tropomyosin: unpublished	60, 61
Vimentin: see ref. 47	68
Keratin: Miles Labs.	69
Nematin: see ref. 36	71
SV40-T: a gift from Dr. G. W. H. Jay, see ref. 33	72
p53: a gift from Dr. G. W. H. Jay, see ref. 10	73
Nuclear Ag: unpublished	74

*In each case the reference referred to can be found in the "References to Chapters 2 and 3" list.

ANTIGEN REFERENCE LIST

Antigen	References*
Actin	16, 24, 31, 43, 58
Alpha-actinin	23, 43
α_2-M	30, 47, 51, 61
Autofluorescence	42, 48
Calmodulin	1, 46, 57
CAT	17
Cathepsin	32
Clathrin	2, 21, 44, 48, 50, 52
EGF-Receptor	4
F_1-ATPase	11, 34
Fibronectin	8, 18, 19, 27, 63, 64
Fodrin	25
Keratin	12, 29
Lipid droplets	56
MC 101	13, 37
Myosin	3, 31, 40, 59
Nematin	36
p21	49, 55
p27	35
p53	10, 20
Pinocytosis	26
Phosphomannosyl receptor (PMR)	54
SV40-T antigen	33
Transferrin receptor (TFR)	53
Tropomyosin	22
Tubulin	6, 9, 14, 28, 38, 41, 47, 48, 60
Vimentin	7, 12
Vinculin	15

*In each case the reference referred to can be found in the "References to Chapters 2 and 3" list.

REFERENCES TO CHAPTERS 2 AND 3

1. Andersen, B., Osborn, M., and Weber, K. (1978). Specific visualization of the distribution of the calcium dependent regulatory protein of cyclic nucleotide phosphodiesterase (modulator protein) in tissue culture cells by immunofluorescence microscopy: Mitosis and intercellular bridge. *Cytobiologie* **17,** 354.

2. Anderson, R. G. W., Vasile, E., Mello, R. J., Brown, M. S., and Goldstein, J. L. (1978). Immunocytochemical visualization of coated pits and vesicles in human fibroblasts: Relation to low density lipoprotein receptor distribution. *Cell* **15,** 919.

3. Ash, J. F., Vogt, P. K., and Singer, S. J. (1976). Reversion from transformed to normal phenotype by inhibition of protein synthesis in rat kidney cells infected with a temperature-sensitive mutant of Rous sarcoma virus. *Proc. Natl. Acad. Sci. U.S.A.* **73,** 3603.

4. Beguinot, L., Lyall, R. M., Willingham, M. C., and Pastan, I. (1984). Down regulation of the epidermal growth factor receptor in KB cells is due to receptor internalization and subsequent degradation in lysosomes. *Proc. Natl. Acad. Sci. U.S.A.* **81,** 2384.

5. Bergmann, J. E., Tokuyasu, K. T., and Singer. S. J. (1981). Passage of an integral membrane protein, the vesicular stomatitis virus glycoprotein, through the Golgi apparatus en route to the plasma membrane. *Proc. Natl. Acad. Sci. U.S.A.* **78,** 1746.

6. Brinkley, B. R., Fuller, G. M., and Highfield, D. P. (1976). Tubulin antibodies as probes for microtubules in dividing and non-dividing mammalian cells. *In* "Cell Motility" (R. D. Goldman, T. D. Pollard, and J. Rosenbaum, eds.), pp. 435–445. Cold Spring Harbor Laboratory, New York.

7. Cabral, F., Willingham, M. C., and Gottesman, M. M. (1980). The localization of an insoluble 58K protein from Chinese hamster ovary cells to 10nm filaments in cultured fibroblasts by electron microscopic immunocytochemistry. *J. Histochem. Cytochem.* **28,** 653.

8. Chen, L. B., Gallimore, P. H., and McDougall, J. K. (1976). Correlation between tumor induction and cell surface LETS protein. *Proc. Natl. Acad. Sci. U.S.A.* **73,** 3570.

9. De Brabander, M., De Mey, J., Joniau, M., and Geuens, S. (1977). Immunocytochemical visualization of microtubules and tubulin at the light and electron microscopic level. *J. Cell Sci.* **28,** 283.

10. Dippold, W. G., Jay, G., DeLeo, A. B., Khoury, G., and Old, L. J. (1981). p53 transformation-related protein: Detection by monoclonal antibody in mouse and human cells. *Proc. Natl. Acad. Sci. U.S.A.* **78,** 1695.

11. Fessenden, J. M., and Racker, E. (1966). Partial resolution of the enzymes catalyzing oxidative phophorylation: XI. Stimulation of oxidative phosphorylation by coupling factors and oligomycin; inhibition by an antibody against coupling factor 1. *J. Biol. Chem.* **241,** 2483.

12. Franke, W. W., Schmid, E., Osborn, M., and Weber, K. (1978). Different intermediate-sized filaments distinguished by immunofluorescence microscopy. *Proc. Natl. Acad. Sci. U.S.A.* **75,** 5034.

13. Fredman, P., Richert, N. D., Magnani, J. L., Willingham, M. C., Pastan, I., and Ginsburg, V. (1983). A monoclonal antibody that precipitates the glycoprotein receptor for epidermal growth factor is directed against the human blood group H type 1 antigen. *J. Biol. Chem.* **258,** 11206.

14. Fuller, G. M., Brinkley, B. R., and Boughter, J. M. (1975). Immunofluorescence of mitotic spindles by using monospecific antibodies against bovine brain tubulin. *Science* **187,** 948.

15. Geiger, B. K., Tokuyasu, T., Dutton, A. H., and Singer, S. J. (1980). Vinculin, an intracellular protein localized at specialized sites where microfilament bundles terminate at cell membranes. *Proc. Natl. Acad. Sci. U.S.A.* **77,** 4127.

16. Goldman, R. D., Lazarides, E., Pollack, R., and Weber, K. (1975). The distribution of actin in nonmuscle cells: The use of actin antibody in the localization of actin within the microfilament bundles of mouse 3T3 cells. *Exp. Cell Res.* **90,** 333.

17. Gorman, C. M., Merlino, G. T., Willingham, M. C., Pastan, I., and Howard, B. H. (1982). The Rous sarcoma virus long terminal repeat is a strong promoter when introduced into a variety of eucaryotic cells by DNA-mediated transfection. *Proc. Natl. Acad. Sci. U.S.A.* **79,** 6777.

18. Hedman, K. (1980). Intracellular localization of fibronectin using immunoperoxidase cytochemistry in light and electron microscopy. *J. Histochem. Cytochem.* **28,** 1233.

19. Hedman, K., Vaheri, A., and Wartiovaara, J. (1978). External fibronectin of cultured human fibroblasts is predominantly a matrix protein. *J. Cell Biol.* **76,** 748.

20. Jay, G., DeLeo, A. B., Appella, E., DuBois, G. C., Law, L. W., Khoury, G., and Old, L. J. (1980). A common transformation-related protein in murine sarcomas and leukemias. *Cold Spring Harbor Symp. Quant. Biol.* **44,** 659.

21. Keen, J. H., Willingham, M. C., and Pastan, I. (1981). Clathrin and coated vesicle proteins: Immunologic characterization. *J. Biol. Chem.* **256,** 2538.

22. Lazarides, E. (1975). Tropomyosin antibody: The specific localization of tropomyosin in nonmuscle cells. *J. Cell Biol.* **65,** 549.

23. Lazarides, E., and Burridge, K. (1975). Alpha-actinin: Immunofluorescent localization of a muscle structural protein in nonmuscle cells. *Cell* **6,** 289.

24. Lazarides, E., and Weber, K. (1974). Actin antibody: The specific visualization of actin filaments in non-muscle cells. *Proc. Natl. Acad. Sci. U.S.A.* **71,** 2268.

25. Levine, J., and Willard, M. (1981). Fodrin: Axonally transported polypeptides associated with the internal periphery of many cells. *J. Cell Biol.* **90,** 631.

26. Lewis, W. H. (1931). Pinocytosis, *Bull. Johns Hopkins Hosp.* **49,** 17.

27. Mautner, V. M., and Hynes, R. O. (1977). Surface distribution of LETS protein in relation to the cytoskeleton of normal and transformed cells. *J. Cell Biol.* **75,** 743.

28. Osborn, M., and Weber, K. (1976). Tubulin-specific antibody and the expression of microtubules in 3T3 cells after attachment to a substratum. *Exp. Cell Res.* **103,** 331.

29. Osborn, M., Franke, W. W., and Weber, K. (1977). Visualization of a system of filaments 7–10 nm thick in cultured cells of an epitheliod line (PtK2) by immunofluorescence microscopy. *Proc. Natl. Acad. Sci. U.S.A.* **74,** 2490.

30. Pastan, I., Willingham, M., Anderson, W., and Gallo, M. (1977). Localization of serum derived alpha$_2$-macroglobulin in cultured cells and decrease after Moloney sarcoma virus transformation. *Cell* **12,** 609.

31. Pollack, R., Osborn, M., and Weber, K. (1975). Patterns of organization of actin and myosin in normal and transformed cultured cells. *Proc. Natl. Acad. Sci. U.S.A.* **72,** 994.

32. Poole, A. R. (1977). Antibodies to enzymes and their uses, with particular reference to lysosomal enzymes. *In* "Lysosomes: a Laboratory Handbook," (J. T. Dingle, ed.), pp. 245–312. Elsevier, Amsterdam.

33. Pope, J. H., and Rowe, W. P. (1964). Detection of specific antigen in SV40-transformed cells by immunofluorescence. *J. Exp. Med.* **120,** 121.

34. Pullman, M. E., Penefsky, H. S., Datta, A., and Racker, E. (1960). Partial resolution of the enzymes catalyzing oxidative phosphorylation: I. Purification and properties of soluble, dinitrophenol-stimulated adenosine triphosphatase. *J. Biol. Chem.* **235,** 3322.

35. Richert, N. D. (1982). Monoclonal antibody specific for avian sarcoma virus structural protein p27. *J. Gen. Virol.* **62,** 385.

36. Richert, N. D., and Willingham, M. C. (1982). Nematin: An antigen present in an unusual crystalloid structure in cultured cells detected by a monoclonal antibody. *J. Cell. Biol.* **95,** 29a.

37. Richert, N. D., Willingham, M. C., and Pastan, I. (1983). Epidermal growth factor receptor: Characterization of a monoclonal antibody specific for the receptor of A431 cells. *J. Biol. Chem.* **258,** 8902.

38. Weber, K. (1976). Visualization of tubulin-containing structures by immunofluorescence microscopy: Cytoplasmic microtubules, mitotic figures and vinblastine-induced paracrystals. *In* "Cell Motility" (R. D. Goldman, T. D. Pollard, and J. Rosenbaum, eds.), pp. 403–418. Cold Spring Harbor Laboratory, New York.

39. Weber, K., Bibring, T., and Osborn, M. (1975). Specific visualization of tubulin-containing structures in tissue culture cells by immunofluorescence. *Exp. Cell Res.* **95,** 111.

40. Weber, K., and Groeschel-Stewart, U. (1974). Antibody to myosin: The specific visualization of myosin-containing filaments in nonmuscle cells. *Proc. Natl. Acad. Sci. U.S.A.* **71,** 4561.

41. Weber, K., Pollack, R., and Bibring, T. (1975). Antibody against tubulin: The specific visualization of cytoplasmic microtubules in tissue culture cells. *Proc. Natl. Acad. Sci. U.S.A.* **72,** 459.

42. Weber, K., Rathke, P. C., and Osborn, M. (1975). Cytoplasmic microtubular images in glutaraldehyde-fixed tissue culture cells by electron microscopy and by immunofluorescence microscopy. *Proc. Natl. Acad. Sci. U.S.A.* **75,** 1820.

43. Wehland, J., Osborn, M., and Weber, K. (1979). Cell-to-substratum contacts in living cells: A direct correlation between interference-reflexion and indirect-immunofluorescence microscopy using antibodies against actin and alpha-actinin. *J. Cell Sci.* **37,** 257.

44. Wehland, J., Willingham, M. C., Dickson, R. B., and Pastan, I. (1981). Microinjection of anti-clathrin antibodies into fibroblasts does not interfere with the receptor-mediated endocytosis of alpha$_2$-macroglobulin. *Cell* **25,** 105.

45. Wehland, J., Willingham, M. C., Gallo, M. G., and Pastan, I. (1982). The morphologic pathway of exocytosis of the G protein of vesicular stomatitis virus in cultured fibroblasts. *Cell* **28,** 831.

46. Welsh, M. J., Dedman, J. H., Brinkley, B. R., and Means, A. R. (1978). Calcium-dependent regulator protein: Localization in mitotic apparatus of eukaryotic cells. *Proc. Natl. Acad. Sci. U.S.A.* **75,** 1867.

47. Willingham, M. C. (1980). Electron microscopic immunocytochemical localization of intracellular antigens in cultured cells: The EGS and ferritin bridge procedures. *Histochem. J.* **12,** 419.

48. Willingham, M. C. (1983). An alternative fixation-processing method for pre-embedding ultrastructural immunocytochemistry of cytoplasmic antigens: The GBS procedure. *J. Histochem. Cytochem.* **31,** 791.

49. Willingham, M. C., Banks-Schlegel, S. P., and Pastan, I. H. (1983). Immunocytochemical localization in normal and transformed human cells in tissue culture using a monoclonal antibody to the *src* protein of the Harvey strain of Murine Sarcoma Virus. *Exp. Cell Res.* **149,** 141.

50. Willingham, M. C., Keen, J. H., and Pastan, I. (1981). Ulstrastructural immunocytochemical localization of clathrin in cultured fibroblasts. *Exp. Cell Res.* **132,** 329.

51. Willingham, M. C., Maxfield, F. R., and Pastan, I. (1980). Receptor-mediated endocytosis of alpha$_2$-macroglobulin in cultured fibroblasts. *J. Histochem. Cytochem.* **28,** 818.

52. Willingham, M. C., and Pastan, I. (1980). The receptosome: An intermediate organelle of receptor-mediated endocytosis in cultured fibroblasts. *Cell* **21,** 67.

53. Willingham, M. C., and Pastan, I. (1985). Ultrastructural immunocytochemical localization of the transferrin receptor using a monoclonal antibody in human cells. *J. Histochem. Cytochem.* **33,** 59.

54. Willingham, M. C., Pastan, I. H., and Sahagian, G. G. (1983). Ultrastructural immunocytochemical localization of the phosphomannosyl receptor in Chinese hamster ovary cells. *J. Histochem. Cytochem.* **31,** 1.

55. Willingham, M. C., Pastan, I., Shih, T., and Scolnick, E. (1980). Localization of the *src* gene product of the Harvey strain of murine sarcoma virus to the plasma membrane of transformed cells. *Cell* **19,** 1005.

56. Willingham, M. C., and Rutherford, A. V. (1984). The use of the osmium-thiocarbohydrazideosmium (OTO) method for the pre-embedding preservation of membranes, and its combination with ferrocyanide-reduced osmium to enhance membrane contrast and preservation in cultured cells. *J. Histochem. Cytochem.* **32,** 455.

57. Willingham, M. C., Wehland, J., Klee, C. B., Richert, N., Rutherford, A. V., and Pastan, I. (1983). Ulstrastructural immunocytochemical localization of calmodulin in cultured cells. *J. Histochem. Cytochem.* **31,** 445.

58. Willingham, M. C., Yamada, S. S., Davies, P. J. A., Rutherford, A. V., Gallo, M. G., and Pastan, I. (1981). The intracellular localization of actin in cultured fibroblasts by electron microscopic immunocytochemistry. *J. Histochem. Cytochem.* **29,** 17.

59. Willingham, M. C., Yamada, S. S., Bechtel, P. J., Rutherford, A. V., and Pastan, I. (1981). Ultrastructural immunocytochemical localization of myosin in cultured fibroblastic cells. *J. Histochem. Cytochem.* **29,** 1289.

60. Willingham, M. C., Yamada, S. S., and Pastan, I. (1980). Ultrastructural localization of tubulin in cultured fibroblasts. *J. Histochem. Cytochem.* **28,** 453.

61. Willingham, M. C., Yamada, S. S., and Pastan, I. (1978). Ultrastructural antibody localization of alpha$_2$-macroglobulin in membrane-limited vesicles in cultured cells. *Proc. Natl. Acad. Sci. U.S.A.* **75,** 4359.

62. Willingham, M. C., Yamada, K. M., Yamada, S. S., Pouyssegur, J., and Pastan, I. (1977). Microfilament bundles and cell shape are related to adhesiveness to substratum and are dissociable from growth control in cultured fibroblasts. *Cell* **10,** 375.

63. Yamada, K. M. (1978). Immunological characterization of a major transformation-sensitive fibroblast cell surface protein. *J. Cell Biol.* **78,** 520.

64. Yamada, S. S., Yamada, K. M., and Willingham, M. C. (1980). Intracellular localization of fibronectin by immuno-electron microscopy. *J. Histochem. Cytochem.* **28,** 953.

INDEX*

*P, plate.